新世纪高职高专行业英语类课程规划教材

建筑实用英语

Architecture Vocationally Related English

主　编　王宝华　张培方
副主编　王　磊　高　馨
编　者　徐晓辉　张香兰　刘延虹　郑素娟
　　　　隋晓辉　毕丽华　张桂兰
主　审　Shannon Arnoldsen(美)　杨峰俊

大连理工大学出版社

图书在版编目(CIP)数据

建筑实用英语 / 王宝华,张培方主编. — 大连:
大连理工大学出版社,2012.7
新世纪高职高专行业英语类课程规划教材
ISBN 978-7-5611-7068-7

Ⅰ.①建… Ⅱ.①王… ②张… Ⅲ.①建筑—英语—
高等职业教育—教材 Ⅳ.①H31

中国版本图书馆CIP数据核字(2012)第146398号

大连理工大学出版社出版

地址:大连市软件园路80号　　　邮政编码:116023
发行:0411-84708842　邮购:0411-84703636　传真:0411-84701466
E-mail:dutp@dutp.cn　　　　　　URL:http://www.dutp.cn
大连印刷三厂印刷　　　大连理工大学出版社发行

幅面尺寸:185mm×260mm　　印张:13.5　　字数:305千字
印数:1～3000
2012年7月第1版　　　　　　2012年7月第1次印刷

责任编辑:张剑宇　　　　　　责任校对:孔　娜
封面设计:张　莹

ISBN 978-7-5611-7068-7　　　　　　定价:32.00元

《建筑实用英语》按照教育部高等教育司颁布的《高职高专教育英语课程教学基本要求》(简称《基本要求》)各项规定及量化指标编写。

本教材适用于三年制建筑工程专业及与建筑工程相关的各专业2+1教学模式的班级。总学时数为64学时。教材基本内容共8课，每课包含5个部分：听说、精读、泛读、语法及应用。

本教材结合高职高专英语教学实际，遵循《基本要求》提出的"立足实用，打好基础，强化能力"的高职英语教学原则。为了适应信息社会、知识经济对现代人的要求，培养学生的综合职业能力和以行动导向学习为目标，根据基础教学向专业化发展的要求，本教材在编写过程中充分注意到：

1. 以素质教育为核心，以培养英语交际能力和阅读能力为重点。本教材的选材充分考虑到英语的工具性和载体作用，广泛搜集建筑人文精神与专业特点相结合的教学素材，其内容涵盖了建筑专业科普知识、历史古迹、中外建筑艺术史上最辉煌的时代，给学生提供了具有时代气息的世界各国建筑业发展新动态与具有实用价值的、可供鉴赏的标志建筑物实物插图，以此来加深学生对建筑专业的了解，激发学生的学习兴趣，培养学生的创作热情。

2. 以行动导向学习为中心，组织各项基本技能训练活动。本教材共编入16篇课文，分8课。我们的编写目标是"学地道英语，看实景插图，查详实资料，供鉴赏收藏"。每篇课文的选材通俗易懂，图文并茂，课文内容突出与网络链接特点，注重培养学生自主学习能力和团队合作意识。我们已经为学习者建立了英语学习网络信息交流服务平台。在每篇课文的练习部分都设置了网络作业，根据课文内容，以命题作文的形式让学生上网查找相关的资料，使学生有机会接触、消化和积累与本专业有关的英语语言现象和建筑领域科技词汇，强化培养语言应用能力。

3. 将教师课堂教学引导模式和学生课下自主学习体验模式相结合，处理好语言知识传授和应用能力培养的关系。不以语法为纲，但不排斥必要的语法知识。为了方便学生自主学习，本教材最后附加了极具参考价值的5项附录。

本教材由第二期"全国高职高专英语类专业教学改革课题"获准立项教材建设研究项目课题组编写，其中每课的听力部分由美国网络在线英语教学专家Shannon Arnoldsen女士编写。建筑工程管理学专家杨峰俊教授和山东省建筑结构专业委员会专家温风军教授对教材的专业术语进行了严格把关。全书承蒙全国建筑工程造价与管理专业委员会专家许传海教授、教育管理学博士生导师申培轩教授和山西大学工程学院英语语言教学博士生导师郑仰成教授审阅并对本教材的编写提出了指导性的建议和建设性的宝贵意见,在此一并致谢。

  本教材在编写过程中得到了山东省教育厅的亲切关心和大力支持,得到了同行专家的指导和帮助,对此我们深表谢忱。同时,我们在编写过程中查阅了大量报纸、杂志、书籍,参考了有关网站的详实资料,在此一并向作者表示诚挚的谢意。

  由于我们的学识水平有限,建筑行业知识及教学经验不足,加之时间仓促,疏漏和不妥之处在所难免,恳请广大教师和学生赐教指正,以期不断修改完善。

  所有意见和建议请发往:dutpwy@163.com
  欢迎访问我们的网站:http://www.dutpbook.com
  联系电话:0411-84707604 84706231

编 者
2012年7月

Warm Regards by an American Education Expert

Shannon Arnoldsen

Teaching English as a foreign language is one of the most daunting tasks an educator can face. I am pleased to have been asked my opinion on how a teacher can best tackle this difficult but rewarding challenge, and I hope my thoughts and insights might be of some assistance to the English teachers at your college.

First, I agree; teachers should seek out new and successful methods, but remember, an unwilling student will not master a subject no matter the method. As you teach, ask yourself, "What makes a willing student?" Each teacher should consider each student's needs and desires as individually as possible. A subject merely thrust upon someone has little opportunity to take root, grow, and produce fruit fit for the benefit of society.

Second, I commend the college for setting goals. Although standards taken to an extreme are undesirable, goals are a motivational tool that can help students and teachers reach further as well as measure their progress.

Finally, I strongly admonish you to seek out ways for your students to converse with native speakers on a regular basis. Research shows that interacting with a native speaker daily will yield the best results. These conversations need not be long; fifteen to twenty minutes a day will suffice. There are many businesses such as mine which work to bring English-learners and English-speakers together through the use of technology.

I wish your college the best in training up the next generation of China. I have a deep respect for your mighty nation and hope to see great success come from your efforts.

Warm Regards,
Shannon Arnoldsen
7/2012
www.iSpeakEnglish.us

Contents

Lesson 1 The Ancient Chinese Architecture

- **Part 1 Listening and Speaking**
  Topic   Architectural Philosophy in China \ 2
- **Part 2 Intensive Reading**
  Text A   The Ancient Chinese Architecture \ 3
- **Part 3 Extensive Reading**
  Text B   The Great Wall \ 8
- **Part 4 Sentence Structures**
  Tenses and Voices \ 11
- **Part 5 Applying**
  Enjoy a Poem and Try to Read It Aloud Creative Writing — the Architectural Philosophy in China \ 19

Lesson 2 The Trend for Constructions in China

- **Part 1 Listening and Speaking**
  Topic   The Rampart Is Gigantic \ 22
- **Part 2 Intensive Reading**
  Text A   Features of Chinese Folk Residences \ 23
- **Part 3 Extensive Reading**
  Text B   Modern Architecture in Beijing, China \ 26
- **Part 4 Sentence Structures**
  Declarative Sentences & Interrogative Sentences \ 31
- **Part 5 Applying**
  Learn More About Zhan Tianyou, the "Father of Chinese Railway" Creative Writing — Features of Building in my Hometown \ 35

Lesson 3 The European Art of Building

- **Part 1 Listening and Speaking**
  Topic   Cooperation on Your Project \ 38
- **Part 2 Intensive Reading**
  Text A   Medieval European Architecture \ 39
- **Part 3 Extensive Reading**
  Text B   Neoclassical Architecture \ 44
- **Part 4 Sentence Structures**
  Imperative Sentence \ 48
  Exclamatory Sentence \ 49
- **Part 5 Applying**
  Enjoy These Sayings and Try to Learn Them by Heart Creative Writing — the Style of Medieval European Architecture \ 50

Lesson 4 The American Architectural Style

- **Part 1 Listening and Speaking**
  Topic   The Pyramids and Temples \ 54
- **Part 2 Intensive Reading**
  Text A   The Development of the American Constructions \ 55
- **Part 3 Extensive Reading**
  Text B   Famous Architects \ 59
- **Part 4 Sentence Structures**
  Sentence Patterns \ 63
- **Part 5 Applying**
  Enjoy These Sayings and Try to Learn Them by Heart Creative Writing — Features of Museums in America \ 67

Lesson 5 The Architectural Art in Asia

- **Part 1 Listening and Speaking**
  Topic   Our New Project \ 70
- **Part 2 Intensive Reading**
  Text A   Potala Palace and Forbidden City \ 71
- **Part 3 Extensive Reading**
  Text B   Buildings in Ancient Kyoto, Japan \ 76
- **Part 4 Sentence Structures**
  Noun Clauses \ 79
- **Part 5 Applying**
  Enjoy the Idioms and Try to Use Them Creative Writing — Features of Temples and Shrines in Japan \ 83

Lesson 6 The Classical Masterpiece in Ancient China

- **Part 1 Listening and Speaking**
  Topic   A Successful Trip \ 86
- **Part 2 Intensive Reading**
  Text A   Irrigation Works and Chain Bridges \ 87
- **Part 3 Extensive Reading**
  Text B   The Summer Palace \ 92
- **Part 4 Sentence Structures**
  Attributive Clause \ 95
- **Part 5 Applying**
  Enjoy These Proverbs and Try to Recite Them Creative Writing — on Zhaozhou Bridge \ 99

Lesson 7 The Ancient Architecture in the World

- **Part 1 Listening and Speaking**
  Topic   Gratitude on a Trip \ 102
- **Part 2 Intensive Reading**
  Text A   The Egyptian Pyramid \ 103
- **Part 3 Extensive Reading**
  Text B   Famous Ancient Relics in the World \ 107
- **Part 4 Sentence Structures**
  Adverbial Clause \ 112
- **Part 5 Applying**
  Enjoy the Famous Sayings and Try to Read Them Aloud Creative Writing — on an Ancient Relic in Egypt \ 116

Lesson 8 Ecology and Architecture

- **Part 1 Listening and Speaking**
  Topic   Asking a Volunteer for Help \ 120
- **Part 2 Intensive Reading**
  Text A   Better City, Better Life \ 121
- **Part 3 Extensive Reading**
  Text B   Konarka Power Plastic \ 126
- **Part 4 Sentence Structures**
  Subjunctive Mood \ 131
- **Part 5 Applying**
  Enjoy the Famous Sayings and Try to Recite Them Creative Writing — on Building Materials \ 136

Appendix

1. Keys to Exercises \ 137
2. Chinese Versions for Text A & B \ 143
3. Glossary \ 155
4. Architectural Technical Terms \ 196
5. Bibliography \ 204

Lesson 1
The Ancient Chinese Architecture

- **Part 1 Listening and Speaking**

  Topic   Architectural Philosophy in China

- **Part 2 Intensive Reading**

  Text A   The Ancient Chinese Architecture

- **Part 3 Extensive Reading**

  Text B   The Great Wall

- **Part 4 Sentence Structures**

  Tenses and Voices

- **Part 5 Applying**

  Enjoy a Poem and Try to Read It Aloud

  Creative Writing — the Architectural Philosophy in China

Lesson 1
The Ancient Chinese Architecture

Part 1 Listening and Speaking

Topic Architectural Philosophy in China

Directions: This part is to train your listening ability. The dialogue will be spoken twice. After the first reading, there will be a pause. During the pause, you must read after it and then listen to it again in order to put the sentences into Chinese.

A: I understand you are studying architecture?

B: Yes. I really enjoy it.

A: What type of architecture is your favorite?

B: Well, I enjoy all types, but my favorite is architecture from the continent of Asia and particularly China.

A: Really? What is the main architectural philosophy in China?

B: Much of Chinese architecture is based on symmetry because it emphasizes balance. At the same time, Chinese architecture also uses curves and engravings on beams and lintels. The details are exquisite.

A: I think many continents have adopted some of the Chinese philosophy. I have seen pavilions supported by pillars in many other countries.

B: The wonderful thing about Chinese architecture is how it is based on ancient traditions, but has still been able to absorb new techniques and modern trends. It's been a pleasure to study and learn more about this fascinating style of architecture.

Chinese Versions

A: _____
B: _____
A: _____
B: _____
A: _____
B: _____
A: _____
B: _____

Part 2 Intensive Reading

Text A

The Ancient Chinese Architecture

As an ancient civilized nation and a great country on the East Asian continent, China possesses a vast territory covering 9.6 million sq. km. and a population accounting for over one-fifth of the world's total, 56 nationalities and a history of 5,000 years, during which it has created a unique, outstanding traditional Chinese architecture that is a particularly beautiful branch in the tree of Chinese civilization.

Design Ideas

All ancient Chinese architecture was built according to strict rules of design that made Chinese buildings follow the ideas of Taoism or other Chinese philosophies. The first designing idea was that buildings should be long and low rather than tall — they should seem almost to be hugging you. The roof would be held up by columns, and not by the walls. The roof should seem to be floating over the ground. The second designing idea was symmetry: both sides of the building should be the same, balanced, just as Taoism emphasized balance. Even as early as the Shang Dynasty, about 1500 BC, Chinese buildings looked pretty much like this, with curved tile roofs and long rows of pillars.

Architecture Features

Chinese architecture constitutes the only system based mainly on wooden structures of unique charming appearance. This differs from all other architectural systems in the world which are based mainly on brick and stone structures. Wooden posts, beams, lintels and joists make up the framework of a house. Walls serve as the separation of rooms without bearing the weight of the whole house, which is unique to China. As a famous saying goes, "Chinese houses will still stand when their walls collapse." The specialty of wood requires antisepsis methods to be adopted, thus develops into Chinese own architectural painting decoration. Colored glaze roofs, windows with exquisite appliqué design and beautiful flower patterns on wooden pillars reflect the high-level of the craftsmen's handicraft and their rich imagination.

Architecture and Culture

Architecture and culture are tightly related to each other. In a sense, architecture is the carrier of culture. Styles of Chinese ancient architecture are rich and varied, such as temples, imperial palaces, altars, pavilions, official residences and folk houses, which greatly reflect Chinese ancient thoughts—the harmonious unity of human beings with nature.

There are two typical types of the Chinese ancient architecture representing the profound influence of Chinese culture, which are Feng Shui and Memorial Arch.

Feng Shui: Chinese traditional theory especially directs the process of architectural construction on the basis of the culture of *The Book of Changes*. Its emphasis is concerned with the harmonious unity of human beings with nature.

Memorial Arch (Paifang): It is the derivative of Chinese feudal society, also called Pailou, unique to China, was built to honor great achievement and virtue of ancestors.

Today, based on its traditional soil, Chinese architecture has absorbed foreign architectural culture and continued to forge ahead by complying with the requirements of our time and using new architectural techniques.

(494 words)

Word Study

Words	Pronunciation	Versions
architecture	['ɑ:kitektʃə] n.	art and science of designing and constructing buildings 建筑学 design or style of a building or buildings 建筑设计（风格）
architectural	[ɑ:ki'tektʃərəl] adj.	of or pertaining to the art and science of architecture 建筑学的，建筑上的
civilize	['sivəlaiz] v.	to educate or improve a person or a society; to make sb's behavior or manners better 教化，使文明，使有教养
civilization	[ˌsivəlai'zeiʃ(ə)n] n.	a society, its culture and its way of life during a particular period of time or in a particular part of the world 社会文明
dynasty	['dinəsti, 'dai-] n.	a series of rulers of a country who all belong to the same family 王朝，朝代
continent	['kɔntinənt] n.	each of the main land masses of the Earth—Europe, Asia, Africa, etc. （地球上的）洲，大陆

philosophy	[fi'lɔsəfi] n.	search for knowledge and understanding of the nature and meaning of the universe and of human life 哲学, 哲学体系
symmetry	['simitri] n.	exact match in size and shape between the two halves of sth. 对称
emphasize	['emfəsaiz] v.	put emphasis on (sth.); give emphasis to (sth.); stress 强调, 重读
curve	[kə:v] n.	line of which no part is straight and which changes direction without angles 曲线, 弧线 v. (cause sth. to) form a curve (使)弯曲
constitute	['kɔnstitju:t] v.	make up or form (a whole); be the components of 组成, 构成(某整体)
beam	[bi:m] n.	long piece of wood, metal, concrete, etc. supported at both ends, that carries the weight of part of a building 梁
lintel	['lintl] n.	horizontal piece of wood or stone over a door or window, forming part of the frame (门或窗的)过梁, 楣
joist	[dʒɔist] n.	one of the long thick pieces of wood or metal that are used to support a floor or ceiling in a building 搁栅, 托梁
collapse	[kə'læps] v./n.	(break into pieces and) fall down suddenly (破碎并)突然倒塌 [sing.] sudden fall; collapsing 突然倒下, 坍塌
antisepsis	[ˌænti'sepsis] n.	(of non-living objects) the state of being free of pathogenic organisms 防腐, 消毒
adopt	[ə'dɔpt] v.	~ sb. (to take sb. into one's family) 收养某人, 过继
reflect	[ri'flekt] v.	to manifest or bring back; to throw or bend back or reflect (from a surface) 反映, 反射
exquisite	['ekskwizit] adj.	extremely beautiful or delicate; finely or skillfully made or done 优美的, 优雅的, 精致的
appliqué	[æˌpli:'kei] n./adj.	a decorative design made of one material sewn over another 嵌花, 贴花, 缝花, 嵌花的, 贴花的
pillar	['pilə] n.	upright column of stone, wood, metal, etc used as a support or an ornament, a monument, etc. 柱子, 支柱
altar	['ɔ:ltə] n.	table or raised flat-topped platform on which offerings are made to a god (供奉神时用以放置供品的)供桌, 祭坛
pavilion	[pə'viljən] n.	light building used as a shelter, e.g. in a park (公园中的)亭子, 阁
harmonious	[hɑ:'məuniəs] adj.	free from disagreement or ill feeling 和谐的, 和睦的, 协调的, 调和的, 音调优美的, 悦耳的

profound	[prə'faund] adj.	deep, intense or far-reaching; very great 深的,深切的,深远的; having or showing great knowledge or insight 知识渊博的
absorb	[əb'sɔ:b] v.	take (sth.) in; suck up 吸收(某事物)
comply	[kəm'plai] v.	act in accordance with someone's rules, commands, or wishes 顺从,答应,遵守
forge	[fɔ:dʒ] v.	(written) to move forward in a steady but powerful way 稳步前进
feature	['fi:tʃə] n.	any of the distinct parts of the face, as the eyes, nose, or mouth 特征,容貌,特写
territory	['teritəri] n.	a region marked off for administration or other purposes; an area of knowledge or interest 领土,范围
column	['kɔləm] n.	a tall solid, vertical post, usually round and made of stone 柱,(通常为)圆形石柱,纪念柱
glaze	[gleiz] v.	to cover something with a glaze, to give it a shiny surface 给……上釉;使光滑,使光亮
tile	[tail] n.	a flat thin slab of fired clay, linoleum, etc. 瓦,瓷砖
brick	[brik] n.	a hard block of baked clay used for building walls, houses, etc. 砖
imperial	[im'piəriəl] adj.	of, relating to, or suggestive of an empire or a sovereign, especially an emperor or empress 帝国的,皇帝的

Proper Nouns

1. Taoism 道教
2. Shang Dynasty 商朝
3. Book of Changes《易经》
4. Memorial Arch (Paifang) 牌坊

Useful Expressions

❶ account for 说明,占,解决,得分
   The cotton accounts for 70% of our export. 棉花占我们出口总量的70%。
❷ according to 依照
   by right; according to law 依据正义;按照法律
❸ rather than 而不是
   rational rather than emotional 理性的而不是感性的
❹ be based on... 建立在……基础上
   based on experience; empirical 根据经验的;全凭观察和实验的

❺ in a sense 在某种意义上
In a sense, your personality lies in your sense of humor. 从某种意义上说,你的人格魅力就在于你的幽默感。

❻ on the basis of 以……为基础
Our trade is conducted on the basis of equality. 我们是在平等的基础上进行贸易。

Notes on Text A

❶ An ancient civilized nation and a great country on the East Asian continent, China possesses a vast territory covering 9.6 million sq. km. and a population accounting for over one-fifth of the world's total, 56 nationalities and a recorded history of 5,000 years, during which it has created a unique, outstanding traditional Chinese architecture that is a particularly beautiful branch in the tree of Chinese civilization.

分析 本句为一个复杂的复合句,句子的主干为 China possesses a vast territory covering 9.6 million square kilometers, a population accounting for over one-fifth of the world's total, 56 nationalities and a recorded history of 5,000 years,谓语动词 possesses 后有 4 个宾语,分别是名词短语 a vast territory covering 9.6 million sq. km., a population accounting for over one-fifth of the world's total, 56 nationalities 和 a history of 5,000 years。An ancient civilized nation and a great country on the East Asian continent 作 China 的同位语;现在分词短语 covering 9.6 million sq. km. 和 accounting for over one-fifth of the world's total 作后置定语分别修饰 territory 和 population;during which it has created a unique, outstanding traditional Chinese architecture that is a particularly beautiful branch in the tree of Chinese civilization 是非限制性定语从句,其中又包含一个由 that 引导的限制性定语从句。

❷ As a famous saying goes, "Chinese houses will still stand when their walls collapse."

分析 as 作连词,引导方式状语从句,表示"如……,像……"。

❸ Today, based on its traditional soil, Chinese architecture has absorbed foreign architectural culture and continued to forge ahead by complying with the requirements of our time and using new architectural techniques.

分析 Chinese architecture has absorbed foreign architectural culture and continued to forge ahead 为此句的主干, has absorbed 为现在完成时;based on its traditional soil 为过去分词短语作状语;by complying with the requirements of our time and using new architectural techniques 为 by 引导的介词短语作方式状语,表示"用某种方法"或"以某种手段"。

Part 3 Extensive Reading

Text B

The Great Wall

The Great Wall of China, one of the greatest wonders of the world, was enlisted in the World Heritage by UNESCO in 1987. It starts at the Jiayu Guan Pass of Gansu Province in the west and ends at the Shanhaiguan Pass of Hebei Province in the east. Just like a gigantic dragon, the Great Wall winds up and down across deserts, grasslands, mountains and plateaus, stretching approximately 6,700 kilometers (4,163 miles) from east to west of China. As one of the Eight Wonders in the world, the Great Wall of China has become the symbol of the Chinese nation and its culture.

The Great Wall is divided into two sections, the east and the west, with Shanxi Province as the dividing line. The west part is a rammed earth construction, about 5.3 meters high on average. In the eastern part, the core of the Wall is rammed earth as well, but the outer shell is reinforced with bricks and rocks. The most imposing and best preserved sections of the Great Wall are at Badaling and Mutianyu, not far from Beijing.

The wall is about 25 feet (7.5 meters) high and 30 feet (9 meters) thick. The top of the wall is paved with brick, forming a road. The road is wide enough to hold ten soldiers marching side by side. There are ramparts, embrasures, peep-holes and apertures for archers on the top, besides gutters with gargoyles to drain rain-water off the parapet walk. Two-storied watch-towers are built at approximately 400-meters internals. The top stories of the watch-tower were designed for observing enemy movements, while the first was used for storing grain, fodder, military equipment and gunpowder as well as for quartering garrison soldiers. The highest watch-tower at Badaling standing on a hill-top is reached only after a steep climb, like "climbing a ladder to heaven". The view from the top is rewarding, however. The Wall follows the contour of mountains that rise one behind the other until they finally fade and merge with distant haze.

There stand 14 major passes (Guan, in Chinese) at places of strategic importance along the Great Wall, the most important being both Shanhaiguan and Jiayuguan. Known as "Tian Xia Di YI Guan" (The First Pass under Heaven), Shanhaiguan Pass is situated between two sheer cliffs forming a neck connecting north China with the northeast. It had been, therefore, a key junction contested by all strategists and many famous battles were fought here. It was the gate of Shanhaiguan that the Ming general Wu Sangui opened to the Manchu army to suppress the

peasant rebellion led by Li Zicheng and so surrendered the whole Ming Empire to the Manchus, leading to the foundation of the Qing Dynasty (1644-1911).

The Great Wall is the largest historical and cultural architecture, and that is why it continues to be so attractive to people all over the world.

(487 words)

Word Study

Words	Pronunciation	Versions
gigantic	[dʒai'gæntik] *adj.*	of very great size or extent; immense 巨大的, 庞大的
approximately	[ə'prɔksimətli] *adv.*	almost correct or accurate, but not completely so; not exact 大约, 近似地
ram	[ræm] *v.*	~ **(against/into)** sth. crash against sth; strike or push sth. with great force 撞击某物, 把地面夯实(如筑路时)
imposing	[im'pəuziŋ] *adj.*	impressive in appearance or manner; grand (外表或举止)壮观的, 令人印象深刻的
rampart	['ræmpɑ:t] *n.*	defensive wall round a fort, etc consisting of a wide bank of earth with a path for walking along the top (宽阔的)防御土墙
embrasure	[im'breiʒə] *n.*	an opening (in a wall or ship or armored vehicle) for firing through (碉堡等)枪眼, 炮眼, [建]斜面墙
aperture	['æpətjuə] *n.*	(formal) a small hole of space in something; narrow opening 窄孔, 隙缝
archer	['ɑ:tʃə] *n.*	person who shoots with a bow and arrows, esp. as a sport or (formerly) in battle 射箭运动员, (旧时)弓箭手
gutter	['gʌtə] *n.*	long (usu. semicircular) metal or plastic channel fixed under the edge of a roof to carry away rain-water 排水檐沟, 天沟
gargoyle	['gɑ:gɔil] *n.*	stone or metal spout in the form of a grotesque human or animal figure, for carrying rain-water away from the roof of a church, etc (作怪异人形或动物形的)滴水嘴, (疏导雨水的)凸饰漏嘴
parapet	['pærəpit] *n.*	(in war) protective bank of earth, stones, etc. along the front edge of a trench 胸墙, 矮护墙, 女儿墙
garrison	['gærisn] *n.*	troops stationed in a town or fort 卫戍部队, 守备部队, 警备部队 *v.* 卫戍部队守备(某地)
rewarding	[ri'wɔ:diŋ] *adj.*	making you feel happy and satisfied because you are doing something useful or important 值得做的, 有益的
contour	['kɔntuə] *n.*	the outer edges of something; the outline of its shape or form 轮廓, 外形 a line on a map that shows points that

		are of equal heights above sea level（地图上表示相同海拔各点的）等高线
strategic	[strə'ti:dʒik] adj.	of strategy; forming part of a plan or scheme 战略（上）的，策略（上）的
strategist	['strætidʒist] n.	someone who is good at planning, especially military movements（尤指军事）战略家
suppress	[sə'pres] v.	to put an end to; prohibit 镇压，抑制，查禁，使止住
surrender	[sə'rendə] v.	stop resisting an enemy, etc.; yield; give up 停止抵抗，投降，屈服，放弃
attractive	[ə'træktiv] adj.	having the power to attract; pleasing or interesting 有吸引力的，诱人的

Proper Nouns

1. UNESCO: 联合国教科文组织，全称如下：United Nations Educational Scientific and Cultural Organization 联合国教育科学及文化组织
2. World Heritage 世界遗产
3. Badaling 八达岭
4. Great Wall at Mutianyu 慕田峪长城
5. Tian Xia Di Yi Guan
   天下第一关，包括山海关城、东罗城以及"天下第一关"城楼、靖边楼、牧营楼、临闾楼等。

Useful Expressions

❶ be divided into 把……分隔成若干部分
The investigators will be divided into three groups. 所有的调查员将被分为三个小组。

❷ as well 倒不如，还是……的好，最好……
It will be as well to stop that young screamer. 最好还是让那个大哭大叫的孩子停下。

❸ far from 远离，远非，远远不，完全不，非但不
He traveled far, strayed far from home. 他到很远的地方旅行；远离家乡。

❹ side by side 并排，并肩
two children walking side by side 两个并肩走路的小孩

❺ as well as 也，又
Action as well as thought is necessary. 行动与思考同样必要。

❻ eight wonders in the world 世界八大奇观

❼ watch-towers 角楼
the lamplight sending from the eight-watch-towers 八角楼的灯光

Notes on Text B

❶ Known as "Tian Xia Di Yi Guan" (The First Pass under Heaven), Shanhaiguan Pass is situated between two sheer cliffs forming a neck connecting north China with the northeast.

分析 过去分词短语 known as "Tian Xia Di Yi Guan" 作状语,现在分词短语 forming a neck connecting north China with the northeast 作伴随状语。

❷ It was the gate of Shanhaiguan that the Ming general Wu Sangui opened to the Manchu army to suppress the peasant rebellion led by Li Zicheng and so surrendered the whole Ming Empire to the Manchus, leading to the foundation of the Qing Dynasty (1644-1911).

分析 It was the gate of Shanhaiguan that the Ming general Wu Sangui opened to the Manchu army to suppress the peasant rebellion led by Li Zicheng 是强调句型,强调宾语;led by Li Zicheng 为过去分词短语作后置定语;... opened to...and so surrendered...为并列谓语;leading to the foundation of the Qing Dynasty 为现在分词短语作伴随状语。

Part 4 Sentence Structures

1. Tenses 时态

时态是谓语动词用来表示动作发生的时间或状态持续的时间的各种形式。

英语时态的种类从时间上看有现在、过去、将来、过去将来之分,从方面上看又有一般、进行、完成、完成进行之分。英语时态共有十六种,现以行为动词 work 为例,列表如下:

时\式	一般式	进行式	完成式	完成进行式
现在时	work / works	am / is / are working	have / has worked	have / has been working
过去时	worked	was / were working	had worked	had been working
将来时	shall / will work	shall / will be working	shall / will have worked	shall / will have been working
过去将来时	should / would work	should / would be working	should / would have worked	should / would have been working

❖ 一般现在时

通常以动词原形表示。主语为第三人称单数时，一般在动词原形后加-s 或-es，-s 或-es 加法和名词复数形式一样。

(1)表示经常性或习惯性的动作，如：

He goes to school by bus. 他乘公交车去上学。

They often go for a walk by the river. 他们经常到河边散步。

(2)表示现在的特征或所处状态，如：

He loves sports. 他热爱体育运动。

They are in the college. 他们在上大学。

(3)表示普遍的真理或永恒的状态，如：

The moon goes round the earth. 月亮绕着地球转。

The sun rises in the east. 太阳从东方升起。

(4)在时间、条件状语从句中，表示将来的动作，如：

As soon as the rain stops, we'll leave. 雨一停，我们就走。

If it is fine tomorrow, we shall go to visit the Great Wall. 如果明天天气好的话，我们要去游览长城。

❖ 现在进行时

现在进行时由助动词 be(am/ is/ are)+现在分词构成。

(1)表示此刻正在进行的动作，如：

Mr. Peter is writing a letter now. 彼得先生正在写信。

The children are having their breakfast. 孩子们正在吃早饭。

(2)表示现阶段正在进行的动作，如：

How are you getting on with your work these days? 这些日子你工作进展如何？

They are preparing the final-term examination now. 他们正在准备期末考试。

❖ 现在完成时

现在完成时由助动词 have/has+过去分词构成。

(1)表示在说话前已经完成的动作，这个动作对现在的情况仍有影响，如：

She has gone to Shanghai. 她去上海了。

We have adopted some of the Chinese architectural philosophy. 我们已经接受了中国建筑理念。

(2)表示过去已经开始，持续到现在，而且还可能持续下去的动作或状态。往往和表示一段时间的状语连用，如：

He has studied architecture for 5 years. 他学习建筑学已经五年了。

He has studied architecture since 1985. 他自从1985年以来一直学习建筑学。

(3) 非延续性动词(如 arrive, come, enter, join, go, die, marry, buy 等)的完成时一般不与 for, since 等表示时间段的状语连用,如：

The young man has joined the army. 那位青年参军了。

She has bought a new house in the countryside. 她在乡下买了一座房子。

(4) 现在完成时不能和表示过去的时间状语连用,但可以和表示不确定时间的状语 already, ever, just, not...yet 等连用,如：

The boat has just arrived at the pier. 船刚到达码头。

We have not yet heard any news from him up till now. 我们到现在也没听到他的音讯。

❖ 现在完成进行时

现在完成进行时由助动词 have/has been +现在分词构成。

(1)表示动作从过去某一时间开始,一直延续到现在,这个动作可能刚刚停止,也可能仍在进行,如：

They have been repairing the Palace Museum. 他们一直在修缮故宫博物院。

The children have been watching television since seven o'clock. 孩子们从七点钟就一直在看电视。

(2)表示一直到说话时为止的一段时间内一再重复的动作,如：

I have been calling you several times in two days. 两天内我给你打了几次电话。

They have been discussing the problem all day long. 他们一整天都在讨论这个问题。

❖ 一般过去时

一般过去时由动词的过去式构成。动词过去式的构成是在动词原形后面加-d 或 ed,不规则动词的过去式,请查阅词典。

(1)表示过去某一时间发生的动作或存在的状态,如：

He worked in an architecture company in 1986. 1986年他曾在一家建筑公司工作。

(2)表示过去经常或反复发生的动作或存在的状态,如：

During my stay in the countryside, I often went hunting. 我住在乡下时经常去打猎。

I often visited the Great Wall when I lived in Beijing. 在北京住的时候,我经常游览长城。

(3)用"used to+动词原形"或用"would+动词原形",也可以表示过去经常或反复发生的动作,如：

Mr. Smith used to take part in shooting competitions. 史密斯先生过去经常参加射击比赛。

Whenever we went in trouble, someone would do us a favor. 每当我们遇到困难,总有人来帮忙。

❖ 过去完成时

过去完成时由助动词had+过去分词构成。助动词had用于各人称。

(1)表示在过去某一时间或动作之前已经发生或完成的动作,如:

Night had fallen by then. 到那时夜幕已经降临。

By the end of last year, we had built many new houses. 到去年年底,我们已经建了许多新房子。

(2)表示过去某一时间开始并一直延续到过去另一时间的动作或状态,如:

Before he retired, he had studied *the Book of Changes* for twelve years. 在退休之前,他研究《易经》已经有十二年的历史了。

❖ 过去将来时

过去将来时由助动词should/would+动词原形构成。第一人称用should,第二、三人称用would。

(1)表示过去预计将要发生的动作或存在的状态。这种时态通常用在宾语从句中,主句的谓语动词是一般过去时,如:

I said on Thursday I should listen to a lecture about Taoism the next day. 星期四我说过我将于第二天去听一场有关道教的讲座。

(2)可以用来表示习惯性的动作,各人称都用would,如:

Whenever I caught a cold, she would send for a doctor. 无论何时我一感冒,她就派人去请医生。

Women would stay at home to do housework in the old days. 旧社会女性要待在家里干家务活。

❖ 过去进行时

过去进行时由助动词be(was/ were)+现在分词构成。

(1) 表示在过去某一时刻或某一段时间正在进行的动作,如:

It was being Shang Dynasty 4000 years ago. 4000年前,这里还处在商朝时期。

The peasant rebellion led by Li Zicheng was fighting against the Ming empire when the Ming general Wu Sangui opened Shanhaiguan Pass to the Manchu army. 李自成农民起义军正在与明王朝激战的时候,明朝将领吴三桂打开山海关引清军入关。

(2)过去进行时可以用来描写故事发生的背景,如:

It was a dark night. The wind was blowing hard and the rain was falling heavily outside. 这是个漆黑的夜晚,外面狂风呼啸,大雨滂沱。

❖ 一般将来时

一般将来时由助动词shall/will+动词原形构成。主语是第一人称,用shall或will,主语是第二、三人称,用will。

(1) 表示将要发生的动作或状态,如:

They will go sightseeing next Monday. 他们下周一去游玩。

(2)表示一种倾向或习惯性动作,如:

We'll die without air or water. 没有空气或水我们就会死。

(3)用"be going to+动词原形"构成将来时,表示不久即将发生的事或最近打算进行的动作,如:

I think it is going to rain. 我看天要下雨了。

(4)用"助动词be+现在分词"构成将来时,表示计划、安排在最近即将发生的动作,这种结构通常只限于动作动词,常用的动作动词有start, come, go, leave等。如:

Hurry up! The train is starting. 快! 火车就要开了。

(5)用"系动词be to+动词原形"构成将来时,表示已经安排、计划或双方协商好了的动作,如:

Forbidden City is to be repaired next month. 紫禁城下个月就准备修缮。

The film is to be shown this weekend. 这部影片本周末就上映。

(6)用"系动词be about to+动词原形"构成将来时,表示近期即将发生的动作,如:

Don't worry, UNESCO is about to send a representative. 别担心,联合国教科文组织马上就派代表来了。

❖ 将来进行时

将来进行时由助动词shall/will be+现在分词构成。

(1)表示将来某一时刻或某段时间正在进行的动作,如:

Tomorrow they will be gone to the village fair. 明天他们去赶集。

(2)表示说话人感到某事即将发生或预计要发生,如:

When I return this time next year, the flowers will be blooming everywhere around the garden. 明年这个时候我再回来时,花园里到处都会鲜花怒放。

❖ 将来完成时

将来完成时由助动词shall/will have +过去分词构成。

主要表示将来某一时刻或将来某一时刻之前完成的动作,这一动作也可能继续进行下去,如:

By seven o'clock this afternoon, we shall (will) have got to Shanghai if the train keeps good line. 如果火车运行正常,我们今天下午七点钟到达上海。

Before bedtime he will have finished his design. 睡觉前，他就能完成他的设计。

❖ **将来完成进行时**

将来完成进行时由助动词 shall/will have been+ 现在分词构成。

表示某个在进行的动作一直持续到将来某一时刻，甚至还会持续下去，如：

By the end of next month, they will have been doing the project for ten years. 到下个月底为止，这项工程他们已经干了十年。

❖ **过去完成进行时**

过去完成进行时由助动词 had been+现在分词构成。

主要表示一直持续到过去某一时刻的动作，该动作可能刚结束，也可能在还进行，如：

My 10-year old son had been taking apart of the old clock and fixing it up again for several times before I came back home last night. 昨晚在我回家之前，我10岁大的儿子已经把这个旧钟表拆卸并重新组装了好几回。

❖ **过去将来进行时**

过去将来进行时由助动词 should/ would be + 现在分词构成。

通常出现在宾语从句中，主句的谓语是过去各时态，从句表示一个在某一时刻正在进行的动作，如：

The government promised that they would be building a new highway here next July. 政府承诺说第二年的7月份将在此修建一条高速公路。

❖ **过去将来完成时**

过去将来完成时由助动词 should /would have + 过去分词构成。

表示从过去某时看来将来某一时刻会业已完成的动作，如：

I thought he would have told you the truth. 我原以为他已经把这件事告诉你了呢。

I believed that by the end of that year the government would have completed the new highway, but I am wrong. 我坚信到那年年底，政府会建好那条高速公路的。但是我错了。

❖ **过去将来完成进行时**

过去将来完成进行时由助动词 should / would have been +现在分词构成。

表示从过去某一时刻看来未来某时之前会一直在进行的动作，如：

She decided to retire that winter. By that time she would have been teaching here for forty years. 她决定那年冬天退休。到那时她就已经在这里任教四十年。

Practice

a. Choose the best answer to complete each sentence.

1) I don't know whether it will rain or not, but if it _____, I shall stay at home.
   A. will   B. rain   C. was   D. does

2) _____ put down the receiver when the telephone rang again.
   A. Scarcely did she   B. Scarcely had she   C. Scarcely she   D. Scarcely she had

3) By the time he takes the final exam, he _____ more than five lectures in this semester.
   A. will attend   B. has attended   C. had attended   D. will have attended

4) The train had already left. If only you _____ five minutes earlier.
   A. are coming   B. had come   C. have come   D. will came

5) At this time tomorrow, they _____ the mountain.
   A. would be climbing B. will be climbing   C. will climb   D. would climb

b. Translate the following sentences into Chinese.

1) The next train leaves at 3 o'clock this afternoon.
2) Are you a student of science in our college now?
3) When you have finished the report, I will have waited for about 3 hours.
4) I had learnt 5,000 words before I entered the university.
5) By the end of last month, we had finished this project according to the contract.
6) We shall send her a glass hand-made craft as her birthday gift.
7) I have been working here for three years.
8) I was on the point of telephoning him when his letter arrived.
9) He is used to swimming in winter.
10) I hope his health will have improved by the time you come back next year.

2. Voices 语态

语态是动词的一种形式，用来说明主语与谓语之间的关系。英语的语态分为两种：主动语态(Active Voice)；被动语态(Passive Voice)。主动语态表示句子的主语是动作的执行者；被动语态表示句子的主语是动作的承受者。

❖ **被动语态的构成形式**

be + done, be 有时态变化；done 代表及物动词和动词短语中的动词过去分词。以动词 do 为例的被动语态变化形式：

时态	主动语态	被动语态
一般现在时	do	am / is /are + done
一般过去时	did	was/were + done
一般将来时	will + do/ be going to+ do	will be + done / be going to be + done
过去将来时	would + do	would be + done
现在进行时	am/is/are + doing	am /is /are being + done
过去进行时	was/were + doing	was/were being + done
现在完成时	has/have + done	has/have been + done
过去完成时	had + done	had been + done
将来完成时	will have + done	will have been+ done

❖ **被动语态的用法**

a) 不知道或者不必要说明执行者是谁。

Potala Palace has been postponed to open till next Friday.

布达拉宫已延期到下周五开放。

b) 动作的承受者是谈话的中心，强调的对象。

The Grand Canal was built in Sui Dynasty.

京杭大运河是在隋朝建成。

c) 有时为了礼貌等原因不愿意说出执行者，或是有意回避。

You are requested to give a performance.

请你给我们表演一个节目。

d) 被动结构能使句子得到更好的安排，承上启下，使其结构更加紧凑。

Princess Wencheng arrived in Tibet and was warmly welcomed by the people there.

文成公主一到西藏就受到当地人民的热烈欢迎。

Practice

Choose the best answer to each sentence.

1) "Today is very cold, isn't it?" "Yes, the river is ____."
   A. freezes   B. freeze   C. frozen   D. freezing

2) He ____ mathematics throughout his college life.
   A. bored   B. bored with   C. was bored   D. was bored with

3) My uncle ____ manager of the firm.
   A. has just made   B. is just being made
   C. has just been made   D. is just made

4) Large sums of money ____ each year in painting the steelwork of bridges.
   A. have spent   B. have to be spent
   C. have to spend   D. spend

5) A candidate for the post ____ at the moment.
   A. is interviewing   B. being interviewing
   C. interviewing   D. is being interviewed

Part 5 Applying

1. Enjoy a Poem and Try to Read It Aloud.

A Red Red Rose

—Robert Burns

O my Luve's like a red, red rose,	哦,我的爱人像红红的玫瑰,
That's newly sprung in June;	在六月里初绽。
O my Luve's like the melodie,	哦,我的爱人像优美的旋律,
That's sweetly play'd in tune.	在甜美音符间回荡!
As fair art thou, my bonnie lass,	美丽如你,我的姑娘,
So deep in luve am I;	我爱你那么深切,
And I will luve thee still, my dear,	亲爱的,我会一直爱你,
Till a' the seas gang dry.	直到大海干涸。
Till a' the seas gang dry, my dear,	亲爱的,直到大海干涸,
And the rocks melt wi' the sun;	到烈阳将岩石熔化,
I will luve thee still, my dear,	亲爱的,我会一直爱你,

While the sands o' life shall run.
And fare thee weel, my only luve,
And fare thee weel a while!
And I will come again, my luve,
Tho' it were ten thousand mile!

只要生命之泉不断。
再见吧,我唯一的爱人,
短暂的别离!
我会再次回来,我的爱人,
即使相隔万里!

作者简介

彭斯(1759–1796)

彭斯是一位杰出的苏格兰民族诗人,他的诗歌强调表达瞬间真实的经历和人的感情,摒除一切概念化的虚饰。彭斯生长在农村家庭,从小就帮忙干农活,见父亲辛勤劳动一生,最后却因过于操劳而死,这使他产生了对当时社会的不满。因此,他在作品中对各种形式的宗教和政治思想给予辛辣的讽刺,而这也成为他的诗歌的特色。他受过很少的教育却博览群书,上知天文地理下知文学典故,然而使他流传于世的却是以民间传说为基础的叙事诗。

2. Creative Writing — the Architectural Philosophy in China

Directions: Please find out one of the representative structures of ancient Chinese architecture online and then describe it with one or two illustrations in terms of its times, features and styles, etc. Print the articles in Paper A4 and send a copy to the teacher's e-mail box.

Lesson 2
The Trend for Constructions in China

- **Part 1 Listening and Speaking**
  Topic   The Rampart Is Gigantic

- **Part 2 Intensive Reading**
  Text A   Features of Chinese Folk Residences

- **Part 3 Extensive Reading**
  Text B   Modern Architecture in Beijing, China

- **Part 4 Sentence Structures**
  Declarative Sentences & Interrogative Sentences

- **Part 5 Applying**
  Learn More About Zhan Tianyou, the "Father of Chinese Railway"
  Creative Writing — Features of Building in my Hometown

Lesson 2
The Trend for Constructions in China

Part 1 Listening and Speaking

Topic The Rampart Is Gigantic

Directions: This part is to train your listening ability. The dialogue will be spoken twice. After the first reading, there will be a pause. During the pause, you must read after it and then listen to it again in order to put the sentences into Chinese.

A: Wow! That rampart is gigantic!
B: Yes it is. It is approximately 25 meters high and 15 meters wide.
A: Why was such an imposing wall built?
B: It protected the city from invaders. Enemy armies had to use a ram and tried to break the wall to enter. See the parapets? They were strategically placed where the garrison could get the best view of enemy soldiers.
A: I bet it would take quite an army to penetrate this wall and force the city withdrawn to surrender. Even though it was used for defense purposes, it is still such an attractive structure!

Chinese Versions

A: _____
B: _____
A: _____
B: _____
_____
_____
A: _____
_____
_____

Part 2 Intensive Reading

Text A

Features of Chinese Folk Residences

Folk residences are the earliest architectural types in the history of China. These residences are usually in diversified forms with no restrictions and built according to local conditions. They are the buildings with the strongest folk styles and local flavors in Chinese architectural art. As a country with vast territory and long history, China has various natural and humanistic environment. As such, residence styles differ from place to place. The main styles include Beijing courtyard houses, cave dwellings on Shaanxi Loess Plateau, stilt houses on steep inclines or projecting over water in west Hunan, Huizhou residences in south Anhui and Hakka earth buildings in Fujian, etc. These age-old types of buildings all have their own unique features.

The beauty of Chinese folk residences lies, above all, in the people-friendliness, with residence spaces and shapes suiting people's living needs; at the same time, built with locally sourced materials, these residences are simple and elegant in color and texture. In addition, the buildings are well arranged, compact and practical.

In terms of the decoration of Chinese folk residences, in most cases, carvings and color paintings are applied to the main structures of the building, with few added parts; key decoration techniques include brick, wood and stone carvings, which are mainly used in the most important and predominant parts like the gate, door and window bars, screen walls, the roof and fire-proof walls. The decoration of folk residences in the south of the Yangtze River and in south Anhui is the most elegant and exquisite, especially the three superb decorating techniques used in houses in south Anhui—brick, wood and stone carvings that are best known for the unparalleled craftsmanship.

Residences in different areas have distinctive local styles and national characteristics. The decoration varies due to the owner's economic status and education. Intellectuals' houses and those of wealthy tycoons are totally different in terms of decoration style and taste. Generally, folk residences are simple and elegant. Many houses have couplets hanging on both sides of the gate. The words and calligraphy of the couplets are both vivid and rustic. The articles and furniture in the hall are elaborate yet practical. Decorative carvings are mostly in quiet colors, giving a graceful feel.

Folk residences not only serve as buildings to live in, but also reflect different customs and tell different life stories. Savoring the art of folk residences is experiencing diversified Chinese ethnic cultures and the history of people's lives.

(408 words)

Stilt Houses

Huizhou Residences

Cave Dwellings

Hakka Earth Buildings in Fujian

Beijing Courtyard Houses

Word Study

Words	Pronunciation	Versions
diversified	[dai'və:sifaid] adj.	having variety of character or form or components; or having increased variety 多样化的
restriction	[ris'trikʃən] n.	a principle that limits the extent of something; an act of limiting or restricting (as by regulation) 限制, 约束
texture	['tekstʃə] n.	the feel of a surface or a fabric; the essential quality of something 质地, 质感
compact	['kɔmpækt] adj.	well constructed; solid; firm 紧凑的, 紧密的, 简洁的
predominant	[pri'dɔminənt] adj.	having superiority in power, influence, etc., over others 主要的, 突出的, 最显著的
superb	[sju:'pə:b] adj.	of surpassing excellence; surpassingly good 宏伟的, 壮丽的, 华美的
intellectual	[inti'lektjuəl] n.	a person who uses the mind creatively 知识分子
tycoon	[tai'ku:n] n.	a very wealthy or powerful businessman 巨富
calligraphy	[kə'ligrəfi] n.	handwriting, esp. beautiful handwriting considered as an art 书法, 笔迹
couplet	[kʌplit] n.	two items of the same kind; a stanza consisting of two successive lines of verse; usually rhymed [pl.]对联
rustic	['rʌstik] n. / adj.	an unsophisticated country person; characteristic of rural life; awkwardly simple and provincial 乡下人, 乡村的

elaborate	[iˈlæbərət] *adj.*	containing a lot of details or parts that are connected with each other 精心制作的,详尽的,复杂的
savor	[ˈseivə] *n. / v.*	the taste experience when a savory condiment is taken into the mouth 味道,气味 *v.* 品尝……滋味
ethnic	[ˈeθnik] *adj.*	connected with a particular race, nation, or tribe and their customs and traditions 种族的
parallel	[ˈpærəlel] *n.*	a relationship or similarity between two things, especially things that exist or happen in different places or at different times 平行

Useful Expressions

1. as such 同样地,照这样
   He is a child, and must be treated as such. 他是个孩子,必须被当作孩子对待。
2. lie in 在于……
   The solution lies in the improvement of the economic environment. 解决办法在于改善经济环境。
3. in terms of 就……而言
   In terms of salary, the job is terrible. 从工资的角度来说,这份工作很糟糕。
4. due to 由于……,归因于……
   The television station apologized for the interference, which was due to bad weather conditions. 电视台为干扰表示歉意,那是由于恶劣的天气状况造成的。

Phrases

1. folk residences 民居
2. Beijing courtyard houses 北京四合院
3. cave dwellings 窑洞
4. Shaanxi Loess Plateau 陕西黄土高原
5. stilt houses 吊脚楼
6. Huizhou residences 徽州民居
7. Hakka earth buildings in Fujian 福建土楼
8. the Yangtze River 扬子江

Notes on Text A

❶ In terms of the decoration of Chinese folk residences, in most cases, carvings and color paintings are applied to the main structures of the building, with few added parts; key decoration techniques include brick, wood and stone carvings, which are mainly used in the most important and predominant parts like the gate, door and window bars, screen walls, the roof and fire-proof walls.

分析 In terms of the decoration of Chinese folk residences 和 with few added parts 都是介词短语作状语；which are mainly used in the most important and predominant parts like the gate, door and window bars, screen walls, the roof and fire-proof walls 为非限制性定语从句，修饰 carvings。

❷ The decoration of folk residences in the south of the Yangtze River and in south Anhui is the most elegant and exquisite, especially the three superb decorating techniques used in houses in south Anhui—brick, wood and stone carvings that are best known for the unparalleled craftsmanship.

分析 especially the three superb decorating techniques 是个省略句型，句尾省略了 are the most elegant and exquisite，用来举例说明扬子江以南和安徽南部的民居装饰；used in houses in south Anhui 为过去分词短语作定语，修饰 the three superb decorating techniques；that are best known for the unparalleled craftsmanship 为限制性定语从句，修饰 brick, wood and stone carvings。

❸ Many houses have couplets hanging on both sides of the gate.

分析 hanging on both sides of the gate 作 couplets 的宾语补足语。

❹ Savoring the art of folk residences is experiencing diversified Chinese ethnic cultures and the history of people's lives.

分析 Savoring 和 experiencing 均为动名词，该动名词及其短语分别作主语和表语。

Part 3 Extensive Reading

Text B

Modern Architecture in Beijing, China

China gave the world the Great Wall, pagoda roofs, screens, joints and other innovations of garden architecture. Yet that was centuries ago. Since then, China has been renowned for its concrete blocks.

Following on the heels of a 1990s building boom, mainland cities are once again in a state of architectural ferment. China is bent on creating a more modern socialist state.

Most of the buildings in Tiananmen Square, beneath the famous portrait of Mao Zedong, are monumental structures, with rigid lines and large columns. They look dynastic, part of post-revolutionary Communist China.

But not far away is a striking new building, the new National Grand Theater, radically different from those in the Square.

The new National Grand Theater is an enormous titanium and glass dome surrounded entirely by water. The entrance is underground. (figure 1)

The building is probably the most controversial structure built in Beijing in the past few years.

"A modern building in an existing place is disturbing. That is a fact," says the designer, French architect Paul Andreu.

Besides Andreu's building, the Chinese government has awarded some of its biggest projects to foreign architects. The Bird's Nest Stadium was designed by Swiss architects, and a Dutchman designed the new headquarters for China's state television network, CCTV.

A mesh of steel bands forms the dome of the National Stadium in Beijing, the Olympic Stadium built for the 2008 Summer Olympics in Beijing, China. (figure 2)

Yet many younger people in Beijing aren't particularly drawn to the flashiness.

"I don't like it," says Luo Qing, 32, who has watched the CCTV towers go up across the street from her office. "I can't deal with such a modern-looking building. I still prefer more traditional Chinese architecture."

Qing's colleague Sun Peng, also 32, agrees.

"I do believe that culture is important," architect Andreu says, standing in front of his egg-shaped theater, what he calls "a cultural island in the middle of a lake."

"I think Beijing should construct buildings that reflect the city's history and culture, especially for important landmarks such as the CCTV tower," Peng says.

But "it seems in many ways to have very little to do with the context of Beijing," says Jeffrey Cody, the Chinese University of Hong Kong professor, who is researching the mainland's architectural history. "Andreu feels that he is being much wronged, but most people feel he has missed the balance between being up-to-date and creating something that resonates with the people."

Designed by the Pritzker Prize-winning Dutch architect Rem Koolhaas, the new CCTV building is one of the largest office buildings in the world. (figure 3)

"Respecting for tradition, Chinese disciplines—such as the relationship between man and nature—can turn Western concepts into something uniquely Chinese", says Ma, who in the early 1980s was a visiting scholar at architect Kenzo Tange's office in Japan. He doesn't worry that Chinese aesthetics will be crowded out in the rush to modernize. The foreign invasion will help locals learn new techniques, he believes, leading to cross-fertilization and, ultimately, the emergence of a modern Chinese architectural style.

"Our traditional culture is so strong, it can never be overpowered," he says.

(527 words)

Word Study

Words	Pronunciation	Versions
pagoda	[pə'gəudə] n.	an Asian temple; usually a pyramidal tower with an upward curving roof 宝塔
joint	[dʒɔint] n.	the point of connection between two bones or elements of a skeleton [建]接点, 接榫
concrete	['kɔnkri:t] n.	a strong hard building material composed of sand and gravel and cement and water 混凝土
ferment	['fə:mənt] n.	commotion; unrest 激动, 纷扰
portrait	['pɔ:trit] n.	a painting of a person's face; a word picture of a person's appearance and character 肖像
monumental	[ˌmɔnju'mentl] adj.	relating or belonging to or serving as a monument; of outstanding significance 丰碑式的, 伟大而不朽的
rigid	['ridʒid] adj.	incapable of or resistant to bending; incapable of compromise or flexibility 严格的, 刚性的
striking	[straikiŋ] adj.	attracting attention; impressive 鲜明的, 引人注目的
controversial	[ˌkɔntrə'və:ʃəl] adj.	marked by or capable of arousing controversy 争论的, 引起争论的
headquarters	[ˌhed'kwɔ:təz] n.	a place from which an organization or a military operation is controlled; the people who work there 总部, 总局
up-to-date	['ʌptə'deit] adj.	reflecting the latest information or changes; in accord with the most fashionable ideas or style 最近的, 更新的
resonate	['rezəneit] v.	sound with resonance; be received or understood 共鸣
discipline	['disiplin] n.	a branch of knowledge; a system of rules of conduct or method of practice 学科, 规律性

aesthetics	[i:s'θetiks] n.	the branch of philosophy that studies the principles of beauty, especially in art 美学
invasion	[in'veiʒən] n.	invading; any entry into an area not previously occupied 入侵
cross-fertilization	[krɔs-ˌfə:tilai'zeiʃən] n.	interchange between different cultures or different ways of thinking that is mutually productive and beneficial 互育
overpower	[ˌəuvə'pauə] v.	overcome by superior force; overcome, as with emotions or perceptual stimuli 击败
construct	[kən'strʌkt] v.	to put together substances or parts, esp. systematically, in order to make or build 构成，建造，施工
architect	['ɑ:kitekt] n.	someone who creates plans to be used in making something (such as buildings) 建筑师，设计师
landmark	['lændmɑ:k] n.	a prominent or well-known object in or feature of a particular landscape 地标,陆标,里程碑
grand	[grænd] adj.	large and impressive in size, scope, or extent; magnificent 主要的,极重要的
dome	[dəum] n.	a concave shape whose distinguishing characteristic is that the concavity faces downward 圆屋顶
innovation	[ˌinəu'veiʃən] n.	a creation (a new device or process) resulting from study and experimentation 创新,革新

Proper Nouns

1. Tiananmen Square 天安门广场
2. Mao Zedong 毛泽东
3. the National Grand Theater 国家大剧院
4. the CCTV Building 中央电视台新楼
5. the Chinese University of Hong Kong 香港中文大学
6. Kenzo Tange 丹下健三，闻名世界的日本建筑师。在五十年代，丹下健三赢得了几乎每一个他参加的竞赛，完成了一系列雄伟的公共建筑与国家事务核心设施等大型设计方案。

Useful Expressions

❶ be renowned for 以……闻名
Suzhou is renowned to the world for its arts and crafts. 苏州以其工艺品闻名全球。

❷ in a state of 处于……状态
in a state of extreme emotion 处于极度热情状态下的

❸ be bent on 一心想要，决心要
He is bent on mastering Spanish. 他决心要掌握西班牙语。

❹ be drawn to 被……所吸引
You will be drawn to the buildings you see while walking along the streets in Shanghai.
走在上海的街道上，你会被两边的建筑物所吸引。

❺ have very little to do with 与……没有多大关系
The heading seemed to have little to do with the text. 这个标题看上去和正文关系不大。

❻ be crowded out 被……排挤
The small shop is crowded out by the big supermarket. 小商店受到大型超级市场排挤。

Notes on Text B

❶ "I do believe that culture is important," architect Andreu says, standing in front of his egg-shaped theater, what he calls "a cultural island in the middle of a lake."

分析 do 用于强调。standing in front of his egg-shaped theater 现在分词短语作伴随状语；what he calls "a cultural island in the middle of a lake" 为同位语从句，作 egg-shaped theater 的同位语。

❷ "Respecting for tradition, Chinese disciplines—such as the relationship between man and nature—can turn Western concepts into something uniquely Chinese," says Ma, who in the early 1980s was a visiting scholar at architect Kenzo Tange's office in Japan.

分析 Respecting for tradition 是现在分词短语，在句中作状语。who 引导非限制性定语从句。

❸ The foreign invasion will help locals learn new techniques, he believes, leading to cross-fertilization and, ultimately, the emergence of a modern Chinese architectural style.

分析 he believes 为插入语，leading to... architectural style 是现在分词短语，在句中作伴随状语。

Part 4 Sentence Structures

Declarative Sentences & Interrogative Sentences 陈述句和疑问句

1. Declarative sentences 陈述句

陈述句:陈述句说明一个事实或者陈述说话人的看法。它的主要功能是传递信息和提供情况。陈述句一般用于直接陈述事实或看法,主语在前,谓语在后。常规情况下陈述句用降调。陈述句有两种:肯定陈述句和否定陈述句。

❖ 肯定陈述句

肯定陈述句表达肯定意义的事实或者看法。它一般通过不同的说话口气和不同的肯定手段(如单词重读、强调词、双重否定、反问句或修辞性问句等)来表达千变万化的肯定意义。有时,说话口气的差异会导致肯定程度的差异,例如:

The feature might be Beijing courtyard houses.(可能性最小)这种特色也许属于北京四合院。

We can't live without discipline or aesthetics. 没有原则和美学我们不能生活。

Nobody can succeed in being an architect without working very hard. 不努力工作没人能成为一个有成就的建筑师。

Isn't it the National Grand Theater? (=It is indeed the National Grand Theater.) 这不是国家大剧院吗?

❖ 否定陈述句

否定陈述句主要用来否定句子的肯定意向或提出对比。否定陈述句按否定范围的大小可分为完全否定和部分否定陈述句。完全否定一般通过 not, no, never, none, nobody, nothing, no one, nowhere 等词表现;部分否定则通过对另外一些词,如:all, both, always, every, everything, everybody, everywhere 等的否定来实现。

I don't like living in cave dwellings. 我不喜欢住在窑洞里。

Nobody saw Kenzo Tange here yesterday. 昨天这儿没人看见丹下健三。

Not everyone likes the architectural style. 不是每个人都喜欢那种建筑风格。

All my friends do not have an apartment. 我有些朋友没有房子。

注意:否定陈述句中有一种情况是要求把从句中的否定转移到主句,这在语法上叫转移否定(transferred negation)。转移否定一般要求不能因为否定的转移而产生意义上的变化或违反惯用法的现象。

I don't believe that stilt houses can be built in the north. 我不相信吊脚楼可以建在北方。

We don't think that they are Huizhou residences. 我们认为这些建筑不是徽州民居。

转移否定适应的动词主要是表示看法和感觉的动词,如:think, believe, suppose, expect, imagine, appear, seem, look as if, feel as if 等。

2. Interrogative sentences 疑问句

疑问句：疑问句用于提出问题。按照句法结构和交际功能，可将疑问句分为四种类型：一般疑问句、特殊疑问句、反意疑问句和选择疑问句。

❖ 一般疑问句

（1）一般疑问句常用来询问一件事或证实一种情况。它经常由一个助动词、情态动词、be 动词或 have 开头，其回答通常是 yes 或 no，或者是含有 yes 或 no 意义的其他词语，因此有些语法书把这种问句也叫 Yes-No Question。口语体中的一般疑问句在没有特殊含义时用升调，书面体中的一般疑问句句末要用问号，如：

—Have you any plan to elaborate? 你有详细的计划吗？
—Yes, I have. 我有。
—Can you write a couplet? 你会写对联吗？
—No, I can't. 不，我不会。

（2）请注意一般疑问句的否定式，答语和汉语有差别，如：

—Can't you answer the question? 难道你不能回答这个问题吗？
—Yes, I can. 不，我能回答。
—Isn't it a wonderful play? 这部戏不好看吗？
—No, it isn't. 是的，不好看。

（3）在肯定回答中，有时可以用 all right, of course, certainly 或 ok 来代替 yes。在否定答语中，为礼貌起见，有时不直接用 no，而用 I'm sorry...，或 I'm afraid...，如：

—May I use your bike? 我可以用你的自行车吗？
—All right. 可以啊。
—Won't you come in and have a drink? 为什么不进来喝一杯呢？
—It's very kind of you, but I'm afraid I have an urgent business to do. 你真好，但我有重要的事要做。

❖ 特殊疑问句

特殊疑问句是对句子中某一特殊部分提出疑问。特殊疑问句以一个疑问代词或疑问副词开头，一般用倒装语序，这和一般疑问句类似。但当主语是疑问代词或由疑问代词修饰时，句子用自然语序，即陈述语序。常见的疑问代词有：who, whose, whom, what, which 等。who 和 what 后有时可加 ever 来加强语气。常见的疑问副词有：where, when, why, how 等。特殊疑问句常用降调，如：

Why did you ask me for a bike? 你为什么跟我借自行车呢？

What does your wife do? 你妻子做什么工作?
How far is it from our school? 那儿离我们学校多远?
Who fell off the bike yesterday? 昨天谁从自行车上摔下来了?
Which dictionary belongs to you? 哪本词典是你的?

特殊疑问句在表达惊奇、愤怒、高兴、悲伤等强烈感情色彩时,可将 ever, on earth, in the world, just, exactly 等词放在疑问代词或疑问副词之后,表示"究竟""到底""准确"等意义,它们可起到强调的作用,使句子的语气得到加强,如:

Who ever won the first place in the contest? 那场比赛中到底是谁赢得了第一名?
What on earth did you see there? 你在那儿到底看见了什么?
Who exactly does it belong to? 这到底是谁的?

在特殊疑问句中,还有一种由特殊疑问词开头的复杂问句,这种特殊疑问句一般用来询问对方对某事的看法、判断、意见、请示等。在这种特殊疑问句中,特殊疑问词在宾语从句中充当不同的语法成分,如:

What do you think has happened to her? (What 作主语)
你想她发生了什么事呢?
Where do you suppose he went? (Where 作状语)
你猜想他去哪儿了?

❖ 选择疑问句

选择疑问句是说话人对问题提供两个或两个以上的答案,供对方选择其中的一种。选择疑问句可供选择的答案之间由 or 连接起来。选择疑问句有两种形式:一般疑问句型的选择疑问句和特殊疑问句型的选择疑问句。

—Which do you like better, this one or that one? 你喜欢哪一个,这个还是那个?
—I like that one.
—Does she like maths or Chinese? 她喜欢数学还是语文?
—Chinese. 语文。
—Shall we wait or go home early? 我们是等还是早点回家?
—Go home early. 早点回家。

❖ 反意疑问句

反意疑问句或称为附加疑问句。反意疑问句是附在陈述句之后,对陈述句所叙述的事实提出反问的一种简短问句。反意疑问句表示怀疑或没有把握。

(1) 陈述句式反意疑问句
陈述部分是肯定的,疑问部分是否定的,如:

— The students are having a class, aren't they? 学生们在上课,不是吗?

— Yes, they are. 对,他们在上课。

— No, they aren't. 不,他们没在上课。

陈述部分是否定的,疑问部分是肯定的,用下列句式:

— Mr. Wang is not at home, is he? 王先生不在家,是吗?

— Yes, he is. 不,他在家。

— No, he isn't. 是的,他不在家。

— Nobody saw you, did they? 没有人看见你,对吧?

— No, they didn't. 是的,没有。

— You've never been to London, have you? 你从没去过伦敦,是吗?

— Yes, I have. 不,我去过。

(2) 祈使句式反意疑问句

如果陈述部分是肯定的,疑问部分可用肯定形式,也可用否定形式,一般常用won't表示邀请,用will, would, can, can't表示告诉别人该做什么,如:

Sit down, won't you? 请坐。

Stop that noise, will you? 别吵了,行不行?

如果陈述部分是否定的,疑问部分可用will you或won't you, 如:

Don't forget it, will (won't) you? 别忘了它,好吗?

Don't talk any more, will (won't) you? 别说话了,行吗?

(3) 感叹式反意疑问句

感叹式的反意疑问句,其疑问部分一般用否定式,如:

What a pretty girl she is, isn't she? 她是个多么漂亮的女孩,不是吗?

How naughty the boy is, isn't he? 这男孩真淘气,不是吗?

Practice

Choose the best answer to complete each sentence.

1) They seldom have lunch at home, _____?
   A. haven't they   B. don't they   C. do they   D. have they

2) The old lady didn't remember _____.
   A. where did she put the key   B. where had she put the key
   C. that where she put the key   D. where she put the key

3) What a hot day, _____?
   A. isn't it   B. is it   C. doesn't it   D. does it

4) Is this _____?
   A. the police are searching for
   B. the police are searching
   C. what the police are searching
   D. what the police are searching for

5) My daughter wants to know something about going to Tokyo for a holiday, _____?
   A. does my daughter   B. does she   C. doesn't she   D. isn't she

6) You can never know what to do next, _____?
   A. can you   B. can't you   C. do you   D. don't you

7) Don't smoke in this area, _____?
   A. will you   B. won't you   C. do you   D. don't you

8) Neither you nor I was asked to show the tickets, _____?
   A. were we   B. was I   C. weren't we   D. wasn't I

9) Peter Buddy likes playing football, but he dislikes playing the piano, _____?
   A. does she   B. doesn't he   C. doesn't she   D. does he

10) The Carpenters used to live in Canada, _____?
    A. usedn't they not   B. didn't they   C. usedn't he   D. didn't he

Part 5 Applying

1. Learn More About Zhan Tianyou, the "Father of Chinese Railway".

Directions: Here is a brief introduction to Zhan Tianyou. Read it carefully, and then tell us a figure that you respect with the aids of the Internet.

Zhan Tianyou (1861-1919), a national hero for his role in building China's railroad system, was born in 1861 in southern China's Guangdong Province and died in 1919. As the chief designer of the first railway project built by Chinese engineers, he was widely regarded as the father of China's railroad.

He was intelligent and interested in machinery when he was a child. At the age of 12, Zhan went to US to study. In 1878, he was admitted to Yale University majoring in civil and railway engineering. As one of the earliest Chinese students studying abroad, he was the first Chinese that became a member of the American Engineers' Association.

After graduating from Yale University in 1881, he returned to China. When construction of the well-known Beijing-Zhangjiakou Railway began in 1905, Zhan Tianyou was appointed as the chief engineer. He succeeded in building the zigzag upwards railway after overcoming the gradient problem by switching back the line. Moreover, he initially used two locomotives instead of one—one pulling, and another pushing the train over the area. The railway was completed in 1909, two years ahead of schedule. It added a brilliant page in the history of Chinese railway construction.

After the Beijing-Zhangjiakou railway was open to traffic, he led a team to extend the railroad to the west of Zhangjiakou. Later, he acted as chief designer in the construction of several other railways in China.

After the 1911 revolution, Zhan Tianyou was appointed chief engineer of the Yue-Han Railway Corporation and built railways from Guangzhou to Shaoguan and from Wuchang to Changsha. In 1913, he became a high official in the Ministry of Communications in the Republic of China and played an important role in setting up the national railway technical standard. In the same year, the China Engineer Association was founded and Zhan Tianyou was elected the chairman. In 1916, he was awarded the doctor's degree of law by Hong Kong University.

2. Creative Writing — Features of Buildings in My Hometown.

Directions: Please write something titled "Features of Buildings in My Hometown" in English with pictures taken by your own or by others. Print the articles in Paper A4 and send a copy to the teacher's e-mail box.

_____
_____
_____
_____
_____
_____
_____
_____
_____

Lesson 3
The European Art of Building

- **Part 1 Listening and Speaking**
  Topic   Cooperation on Your Project

- **Part 2 Intensive Reading**
  Text A   Medieval European Architecture

- **Part 3 Extensive Reading**
  Text B   Neoclassical Architecture

- **Part 4 Sentence Structures**
  Imperative Sentence
  Exclamatory Sentence

- **Part 5 Applying**
  Enjoy These Sayings and Try to Learn Them by Heart
  Creative Writing — the Style of Medieval European Architecture

Lesson 3
The European Art of Building

Part 1 Listening and Speaking

Topic Cooperation on Your Project

Directions: This part is to train your listening ability. The dialogue will be spoken twice. After the first reading, there will be a pause. During the pause, you must read after it and then listen to it again in order to put the sentences into Chinese.

A: Hello Mr. Johnson. My name is Steven. I'm one of the architects who will be working on your project. It's very nice to meet you.
B: Thank you. Please call me Mark. I'm excited to get this project going. I hired your company because of the prestigious work you've done downtown.
A: Well, I'm glad we have a good reputation. OK. I understand you want a small pagoda built here in your garden.
B: That's right. I want it to be a real showpiece, something that will really resonate with the landscaping.
A: Let's go over the blueprints together. It shows the foundation and the pillars being made of concrete. Is that correct?
B: Yes. I wanted the structure to be rigid. There's nothing more disturbing than a building that seems weak. And I don't want to worry about a termite invasion. What do you think about the roof?
A: I think it's striking, and because it is made of the most up-to-date materials available, it should last a long time.
B: Yes, that is ultimately my desire. I want a powerful, dynastic look without it overpowering the entire garden.
A: Very good. We'll get right to work.

Chinese Versions

A: _____
B: _____
A: _____
B: _____
A: _____
B: _____
A: _____
B: _____
A: _____

Part 2 Intensive Reading

Text A

Medieval European Architecture

In the medieval period, many different styles of architecture developed. Churches that tried to imitate Roman vaults and arches are now called 'Romanesque'. Later churches that used pointed arches and more decorative window tracery are called 'Gothic'. Gothic is subdivided into different local styles. In England we find the Early English, Decorated, and Perpendicular styles. In France we find High Gothic, Flamboyant, and Rayonnant. The names of local styles are mostly derived from ideas that describe the effect of the architecture, or the shapes of tracery.

'Romanesque' also refers us back to the form of the building, which is reminiscent of those of Rome. In England and northern France, Romanesque architecture can be called 'Norman', after the dukes of Normandy, similarly the name 'Gothic' is political, in which it refers to a body of people, the Goths, who laid waste to Rome. This does not tell us anything useful about the architecture, but it does tell us about how the architecture was seen at the time when the name was coined, in the 17th century, long after the cathedral-builders had stopped building in that manner.

The most spectacular monuments of the Middle Ages are the great French cathedrals, such as Bourges, which were made so as to appear as if they were constructed out of little more than colored light. Cathedral of St. Etienne, Bourges, France began in 1190.

Of all the medieval cathedrals, this is the one that best illustrates the idea of a building as a cage of light. The west end of the building is immensely solid, with five unequal doorways surrounded by hundreds of small-sculpted figures. Above there are two unequal towers. The rest of the building, however, gives an impression of being precisely repetitive, as a standard type of bay, is taken along the whole length of the building without interruption, and adapted with a minimum of difference so as to make it turn the semi-circular end at the west. The nave is immensely high, at 39 meters (125feet), and it is flanked by double aisles. The building is filled with light that filters through stained glass, painted with images of biblical stories. From the outside the buttress that holds the building up is clearly visible, looking like a series of powerful props: they shore up the illusion of weightless delicacy within.

(384 words)

Word Study

Words	Pronunciation	Versions
medieval	[ˌmedi'i:vl] adj.	mediaeval of, relating to, or in the style of the Middle Ages; old-fashioned; primitive 中古的, 中世纪的

vault	[vɔ:lt] n.	an arched structure that forms a roof or ceiling 拱形圆屋顶,穹窿
decorative	['dekərətiv] adj.	serving to decorate or adorn; ornamental 装饰的,装潢的,可作装饰的
tracery	['treisəri] n.	a pattern of interlacing ribs, esp. as used in the upper part of a Gothic window, etc. 由线纹构成的装饰图样
subdivide	[ˌsʌbdi'vaid] v.	to divide (something) resulting from an earlier division 再分,细分
reminiscent	[ˌremi'nis(ə)nt] adj.	stimulating memories (of) or comparisons (with) 使人联想起,怀旧的,回忆的
cathedral	[kə'θi:drəl] n.	the principal church of a diocese, containing the bishop's official throne (as modifier) 大教堂
spectacular	[spek'tækjulə] adj.	of or resembling a spectacle; impressive, grand, or dramatic 引人注目的,壮观的
monument	['mɔnjumənt] n.	a notable building or site, especially one preserved as public property 纪念碑,纪念性建筑物
illustrate	['iləstreit] v.	to clarify or explain by use of examples, analogy, etc 举例说明,例证
sculpt	[skʌlpt] v.	to make a particular shape from wood, stone, clay etc 雕刻 n. sculpture adj. sculpted
repetitive	[ri'petitiv] adj.	repetitious characterized by or given to unnecessary repetition 重复的,反复性的
interruption	[ˌintə'rʌpʃən] n.	something that interrupts; an interval or intermission 中断,打断,障碍物,遮断物
nave	[neiv] n.	central space in a church, extending from the narthex to the chancel and often flanked by aisles (教堂的)正殿
flank	[flæŋk] n.	the side of building 侧面,厢房 v. to be on both sides of someone or something 侧面与……相接
aisle	[ail] n.	a passageway separating seating areas in a theatre, church, etc; gangway(教堂的)走廊,侧廊,耳堂
filter	['filtə] v.	to pass (through a filter or something like a filter) 过滤,渗透
buttress	['bʌtris] n.	also called pier, a construction, usually of brick or stone, built to support a wall 撑墙,扶壁,(前)扶垛,支[肋]墩
delicacy	['delikəsi] n.	fine or subtle quality, character, construction, etc 娇嫩,优美,精致

Useful Expressions

❶ derive from 衍生,起源于……
Many English words derive from France. 很多英语单词源自法语。

❷ refer to 查阅,提到,谈到
The new law does not refer to land used for farming. 这条新的法律不适用于农用土地。

❸ as if 好像,仿佛
He acts as if he knew me. 他好像觉得认识我。

❹ take along 随身带着,随……带来
You'd better take your umbrella along. 你最好带上雨伞。

❺ fill with 充满,用……填充
Fill the bottle with water. 把瓶子装满水。

❻ look like 像……似的
What do photographs look like? 这些照片照得怎样?

Architectural Terms

❶ Romanesque [ˌrəumə'nesk] 【建】罗马式建筑; 罗马式绘画 [雕刻]
❷ Gothic ['gɔθik] 【建】哥特式的 (尖拱式建筑)
❸ Perpendicular [ˌpəːpən'dikjulə] 【建】垂直式的
❹ Flamboyant [flæm'bɔiənt] 【建】火焰式的
❺ Rayonnant ['reiənənt] 【建】(窗格等) 幅射式的,光芒四射的,明亮照耀的
❻ Bourges 布尔日,位于奥尔良东南偏南地区,法国中部城市,曾是奥古斯都时期的罗马首府。

Notes on Text A

❶ 'Romanesque' also refers us back to the form of the building, which is reminiscent of those of Rome.

   分析 refer to "涉及……,谈到……"; 此句的"refer sb back to sth"意思是"将某人带回到……"; which引导非限制性定语从句。

❷ In England and northern France, Romanesque architecture can be called "Norman", after the dukes of Normandy, similarly the name 'Gothic' is political, in which it refers to a body of people, the Goths, who laid waste to Rome.

   分析 in which ... to Rome 是"介词+关系代词"引导的非限制性定语从句,修饰先行词 England and northern France; 该从句中又含有一个由who引导的非限制性定语从句,修饰先行词 people。

❸ This does not tell us anything useful about the architecture, but it does tell us about how the architecture was seen at the time when the name was coined, in the 17th century, long after the cathedral-builders had stopped building in that manner.

分析 but it does tell us ...中的 does 是强调谓语动词 tell 的助动词。...when the name was coined 为限制性定语从句,修饰先行词 time,when 是关系副词;coin 是动词,意为"杜撰"; in that manner 表示"用那种方式"。

❹ The most spectacular monuments of the Middle Ages are the great French cathedrals, such as Bourges, which were made so as to appear as if they were constructed out of little more than colored light.

分析 which 是关系代词,引导后面的非限制性定语从句,修饰先行词 cathedrals ; so as to "如此……以至于,使得"; as if 是连接词引导方式状语从句。

❺ The rest of the building, however, gives an impression of being precisely repetitive, as a standard type of bay, is taken along the whole length of the building without interruption, and adapted with a minimum of difference so as to make it turn the semi-circular end at the west.

分析 gives an impression of "给人……的印象"; as a standard type of bay 是介词短语,在句中作方式状语; a minimum of "最少的……"; 全句为简单句; gives..., is taken..., and adapted... 是并列谓语。

Images Connected with Internet 欣赏相关视频截图

Figure 1-3 巴黎圣母院 风格:早期哥特式 关键词:划时代

巴黎圣母院兴建于12世纪至14世纪,是欧洲早期哥特式建筑的主要代表,有作家形容它重如大象,轻如飞蛾,是法国建筑史上的杰作。正面一对60余米高的塔楼,巍峨而壮观,上面有内容丰富、精美绝伦的反映宗教题材的雕刻。走进大教堂光线骤然暗淡下来,教堂内高大宽敞而幽暗,顿时让人产生一种庄严肃穆的感觉,仿佛进入一个与世隔绝的境地。

作为欧洲建筑史上一个划时代的标志,巴黎圣母院的正外立面风格独特,结构严谨,看上去十分雄伟庄严。它被壁柱纵向分隔为三大块;三条装饰带又将它横向划分为三部分,其中,最下面有三个内凹的门洞。而教堂内部则极为朴素,几乎没有什么装饰。

圣玛丽亚鲜花大教堂是文艺复兴时期第一座伟大建筑。1295年由阿尔诺沃·迪卡姆比奥在原先的佛罗蒂诺大教堂的基址上兴建,1496年才最后完工。

这座教堂大圆顶是世界上第一座大圆顶,是菲利浦·布鲁内莱斯基(1377~1446)的杰作,设计并建造于1420年到1434年间,这位巨匠在完成这一空中巨构的过程中没有借助於拱架,而是用了一种新颖的相连的鱼骨结构和以橡固瓦的方法从下往上逐次砌成。圆顶呈双层薄壳形。

双层之间留有空隙,上端略呈尖形。它高91米,最大直径45.52米。圆顶内部原设计不作任何装饰,后来瓦萨里和祖卡里(1572~1579)在里面画了壁

画。屋顶灯亭也是由布鲁内莱斯基设计的。连灯亭在内,教堂总高为107米。圆顶内有螺旋形阶梯直通穹顶,可鸟瞰全市风光。圆顶内还陈列了米开朗杰罗雕刻的圣彼德像和约2千平方米的巨幅壁画《末日的审判》。

Figure 4 意大利米兰大教堂 风格:以哥特式为主 关键词:多元化

米兰大教堂是意大利米兰的重要地标。整座建筑物融合哥特、文艺复兴、新古典等多种建筑风格,其中教堂正面以三角形状构建而成,四周有数百座尖塔成林,和高达108公尺的主塔相衬,仿如相互扶持,伸向蓝天,展现了哥特式建筑的特色。

对哥特式建筑迷而言,它是装饰精致的杰作:"用大理石写成的一首诗"。仅教堂外部就用了2245尊雕塑、135个尖顶、96个怪兽状滴水嘴和约1公里长的窗花格。不管从哪一个方向看,米兰大教堂呈现的壮丽风貌及丰富的历史内涵,都很令人感动。

Figure 5-6 德国北莱茵－威斯特清伦州科隆大教堂 风格:中晚期哥特式 关键词:最完美

素有欧洲最高尖塔之称的科隆大教堂,建在莱茵河畔,是中晚期哥特式建筑的典范。1996年联合国教科文组织将其作为文化遗产,列入《世界遗产名录》。

据说,科隆大教堂是最完美的哥特式大教堂,它始建于1248年左右,建筑面积约6000平方米。整个建筑建造前后整整持续了632年,是欧洲建筑史上建造时间最漫长的建筑物之一。

雄伟的哥特式建筑,有轻盈雅致的教堂内景,教堂中央是两座与门墙连砌在一起的双尖塔,高161米,是全欧洲最高的尖塔,四周林立着的无数座小尖塔与双尖塔相呼应。教堂内有10座礼拜堂,中央大礼拜堂穹顶高43米,中厅部跨度为15.5米,是目前尚存的最高的中厅。

古典复兴建筑是18世纪60年代到19世纪流行于欧美一些国家的,采用严谨的古希腊、古罗马形式的建筑,又称新古典主义建筑。

当时,人们受启蒙运动的思想影响,崇尚古代希腊、罗马文化。在建筑方面,古罗马的广场、凯旋门和记功柱等纪念性建筑成为效法的榜样。当时的考古学取得了很多的成绩,古希腊、罗马建筑艺术珍品大量出土,为这种思想的实现提供了良好的条件。

采用古典复兴建筑风格的主要是国会、法院、银行、交易所、博物馆、剧院等公共建筑和一些纪念性建筑。这种建筑风格对一般的住宅、教堂、学校等影响不大。

Figure 7 当代建筑新视野

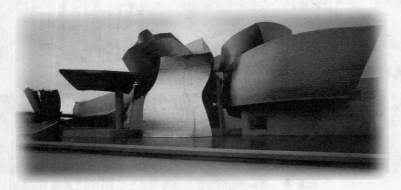

Part 3 Extensive Reading

Text B

Neoclassical Architecture

Neoclassical architecture was an architectural style produced by the neoclassical movement that began in the mid-18th century, both as a reaction against the Rococo style of anti-tectonic naturalistic ornament, and an outgrowth of some classicizing features of Late Baroque. This movement was also nourished with new knowledge from archaeological investigations in Greece, which brought to light a better understanding of the architectural forms of the ancient Greek world.

This architectural style was mainly used in building banks, museums, theaters, and other types of public buildings and commemorative structures. It had little influence on the design of homes, churches or schools.

In the late-18th century to the early-19th century, France was the center of neoclassical architecture in Europe. The Pantheon in Paris is a typical example of classical architecture during the French Revolution. In Napoleon's times there were many commemorative structures being built in Paris. One of them is the Arc de Triomphe, which imitates the type of ancient Roman architecture. The Arc de Triomphe (figure 8) is a monument in Paris, France that stands in the centre of the Place Charles de Gaulle, also known as the "Place de l'Étoile". It is at the western end of the Champs-Élysées. The triumphal arch honors those who fought for France, particularly during the Napoleonic Wars. The monument stands 49.5 meters (162 feet) in height, 45 meters (150 feet) wide and 22 meters (72 feet) deep. It is the second largest triumphal arch in existence. Its design was inspired by the Roman Arch of Titus (figure 9).

9

By the end of the 18th century, then, their rival versions of classicism are in circulation, based on various understandings of Greek and Roman architecture, ranging from the fundamentalist simplicity of austere Doric temples, to the highly ornamental work of the Adam brothers. There was also a growing antiquarian interest in medieval architecture of the Gothic Revival of the mid-19th century and various forays into exotic spectacle, such as the Royal Pavilion. This eclecticism has flourished ever since, more visibly at some times than at others, marking the fact that the tastes of the classes that have money to spend on building were no longer unified.

(346 words)

Word Study

Words	Pronunciation	Versions
tectonic	[tek'tɔnik] *adj.*	relating to construction or building 构造的, 建筑的
naturalistic	[ˌnætʃərə'listik] *adj.*	a naturalistic style 自然的, 自然主义的
ornament	['ɔ:nəmənt] *n.*	decoration; adornment 装饰, 修饰 ornamental [ˌɔ:nə'mentl] *adj.* 作装饰用的
outgrowth	['autgrəuθ] *n.*	a development, result, or consequence 自然的发展, 自然结果
nourish	['nʌriʃ] *v.*	to support or encourage (an idea, feeling, etc) 滋养, 养育
archaeological	[ˌɑ:kiə'lɔdʒikəl] *adj.*	related to or dealing with or devoted to archaeology 考古学的, 考古学上的
investigation	[inˌvesti'geiʃən] *n.*	a careful search or examination in order to discover facts, etc. 调查, 调查研究
commemorative	[kə'memərətiv] *adj.*	intended as a commemoration 纪念的
rival	['raivəl] *n.*	a person, a team, etc. that competes with another for the same object or in the same field 对手, 竞争者
circulation	[ˌsə:kju'leiʃən] *n.*	passing of something from one person or place to another; spread 流通, 传播
austere	[ɔs'tiə] *adj.*	severely simple or plain, serious 严峻的, 苛刻的
antiquarian	[ˌænti'kwɛəriən] *adj.*	concerned with the study of antiquities or antiques 研究文物的, 收藏古物的
foray	['fɔrei] *n.*	brief but vigorous attempt to be involved in a different activity (对一新的活动) 短暂而积极的改换尝试

exotic	[ig'zɔtik] adj.	having a strange or bizarre allure, beauty, or quality 式样奇特的,异国情调的
spectacle	['spektəkl] n.	a strange or interesting object or phenomenon (引人注目的) 景象,奇观
flourish	['flʌriʃ] v.	very active, or widespread; prosper 繁荣,兴旺,活跃,盛行
revival	[ri'vaivəl] n.	bringing again into activity and prominence 苏醒,复活,再生

Proper Nouns

1. Rococo [rə'kəukəu] n. 洛可可风格起源于18世纪的法国,最初是为了反对宫廷的繁文缛节艺术而兴起的。洛可可 Rococo 这个字是从法文 Rocaille 和意大利文 Barocco 合并而来。Rocaille 是一种混合贝壳与小石子制成的室内装饰物。洛可可后来被新古典主义取代。

2. Baroque [bə'rəuk] 巴罗克艺术风格(17世纪初流行于欧洲的一种过分强调雕琢和装饰奇异的艺术和建筑风格,倾向于豪华、浮夸,并将建筑、绘画、雕塑结合成一个整体,追求动势的起伏,以求造成幻象的建筑形式);变态式。

3. Pantheon [pæn'θi(:)ən]【史】(古希腊、罗马供奉众神的)万神殿(罗马一圆顶庙宇,建于公元120—124)

4. Napoleon [nə'pəuljən,-liən] 拿破仑(Napoleon I 拿破仑一世,全称 Napoleon Bonaparte, 1769—1821 法国皇帝,1804—1815在位)

5. Arc de Triomphe (法语),英文表达凯旋门 (triumphal arch),1806年由拿破仑下令兴建,用于纪念法军在奥斯特里茨战役中战胜俄奥联军。全部工程用了30年时间建成,于1836年即拿破仑去世15年后才落成。后来成为法国的无名英雄墓,纪念一次大战中阵亡的150万法国士兵。

6. Place Charles de Gaulle (法语)戴高乐广场,该广场原名星形广场(Place de l'Étoile)

7. Champs-Élysées 香榭丽舍大街(法语:Avenue des Champs-Élysées 或 les Champs-Élysées)是巴黎一条著名的大街,位于城市西北部的第八区。它被誉为巴黎最美丽的街道。

8. Napoleonic Wars 拿破仑战争是指拿破仑称帝统治法国期间(1804—1815)爆发的各场战事,这些战斗可说是自1789年法国大革命所引发的战争的延续。它促使了欧洲的军队和火炮发生重大变革,特别是军事制度。因为实施全民征兵制,使得战争规模庞大,史无前例。法国国势迅速崛起,雄霸欧洲,但是在侵俄战役惨败后,一落千丈。拿破仑建立的帝国,最终战败,让波旁王朝得于1814年和1815年两度复辟。

9. Roman Arch of Titus 提图斯凯旋门是意大利罗马市古罗马广场东南圣道上的一座大理石单拱凯旋门,由图密善皇帝兴建于兄长提图斯去世后不久,纪念在公元70年征服和摧毁耶路撒冷,终止66年开始的犹太人大起义。提图斯凯旋门是16世纪以后许多凯旋门仿效的对象。

10. Fundamentalist [fʌndə'mentəlist] 原教旨主义者,信奉正统派基督教的人

11. Doric ['dɔrik]【建】陶立克式的(纯朴、古老的希腊建筑风格)

12. Adam brothers 指18世纪英国著名建筑师罗伯特·亚当与詹姆斯·亚当兄弟二人,他们主张一种"尚希腊"风格但具英格兰人欣赏趣味的建筑。

13. Gothic Revival 哥特式艺术,又译作哥德式艺术,为一种源自欧洲法国的艺术风格,该风格始于12世纪的法国,盛行于13世纪,至14世纪末期,其风格逐渐大众化和自然化,成为国际哥特风格,直至15世纪,因为欧洲文艺复兴时代来临而迅速没落。不过,在北欧地区,这种风格仍延续相当长的一段时间。该风格在18世纪重新被肯定,"哥特复兴"(Gothic Revival)运动推崇中世纪的阴暗情调,在19世纪之后仍偶尔被应用。

14. Royal Pavilion 皇家穹顶宫是位于英国海滨旅游胜地布赖顿的豪华宫殿,在19世纪摄政王(后来的英国国王乔治四世)的海边隐居地。1783年,摄政王首次访问布赖顿,进行海水物理治疗痛风。1786年他租下一间农舍。1815—1822,建筑师约翰·纳西(John Nash)重新设计宫殿,今天看见的就是他的作品。宫殿就在布赖顿的中心。外观受到印度伊斯兰建筑风格(莫卧儿王朝)的强烈影响,有点类似泰姬陵。富于幻想的内部设计,基本上出自弗雷德里克·格雷斯(Frederick Crace)和罗伯特·琼斯(Robert Jones)的设计思路,内部装饰和摆设则充满中国情调。这是个完美异国情调的例证,是对摄政风格更加古典的主流口味的一种变通办法。

15. eclecticism [e'klektisizəm] 折衷主义,折衷主义的运动

Useful Expressions

1. influence on 对……有影响
   This book has great influence on the readers. 这本书对读者产生了长远影响。
2. fight for 为……而战
   They fight for defending the country. 他们为保卫祖国而战斗。
3. base on 以……为基础；基于……
   A good marriage is based on trust. 美满的婚姻是建立在互相信任的基础上的。
4. range from... to 范围从……到……
   Their ages range from 25 to 50. 他们的年龄在25岁到50岁之间。

Notes on Text B

1. Neoclassical architecture was an architectural style produced by the neoclassical movement that began in the mid-18th century, both as a reaction against the Rococo style of anti-tectonic naturalistic ornament, and an outgrowth of some classicizing features of Late Baroque.

   分析 that引导限定性定语从句。both ... and "两者都……"，and后面省略了as，该并列连词连接的是两个由as引导的介词短语，修饰全句，作方式状语。

2. The Arc de Triomphe is a monument in Paris, France that stands in the centre of the Place Charles de Gaulle, also known as the "Place de l'Étoile".

   分析 in the centre of "在……的中心"；known as "被认为是……"。

3. By the end of the 18th century, then, their rival versions of classicism are in circulation, based on various understandings of Greek and Roman architecture, ranging from the fundamentalist simplicity of austere Doric temples, to the highly ornamental work of the Adam brothers.

   分析 based on... architecture 过去分词短语，在句中作非限制性定语，修饰versions；ranging from... to 现在分词短语，在句中作伴随状语。

4. This eclecticism has flourished ever since, more visibly at some times than at others, marking the fact that the tastes of the classes that have money to spend on building were no longer unified.

   分析 ever since "从那时起"。the fact that... 其中"that"是连接代词，引导同位语从句。the classes that... 其中"that"是关系代词，在从句中作主语，修饰先行词classes，引导限制性定语从句；no longer "不再"。

Part 4 Sentence Structures

1. Imperative Sentence 祈使句

用以表示请求、命令、劝告、建议等的句子叫祈使句。祈使句的结构与陈述句一样，但主语常被省略。祈使句一般没有时态变化，也不能与情态动词连用。祈使句的主语通常为第二人称you，谓语动词用动词原形，句尾用句号或感叹号。否定结构用do not 或其缩略式

don't 加动词原形。

❖ 构成形式

Take Baroque for example. 以巴罗克艺术风格为例。
Come and help! 来帮个忙吧!
Don't worry about the rival, Walt. 别担心竞争对手,沃尔特。
Don't be so austere! 不要如此苛刻!
Don't forget neoclassical architecture! 不要忘记新古典主义建筑风格!

(1) 祈使句用于第一人称则使用谓语动词 let,let me 表单数,let us (let's) 表复数,如:
Let me tell you what Gothic Revival is. 我告诉你什么是哥特式建筑风格吧。
Let us be calm, gentlemen. 大家镇定,先生们。
Let's go on talking about the history of Rococo. 我们继续说说洛可可的历史。

(2) 否定结构通常将否定词 not 置于 let me 或 let us (let's) 之后,如:
Let's not waste our time arguing about if the new law refers to the farming land.
咱们别浪费时间争论这项新法律是否适用于农用土地了。
Let's not say anything about the build method.
关于这种建筑方法咱们谁也别谈论了。

2. Exclamatory Sentence 感叹句

表示说话时的惊异、喜悦、气愤等情绪的句子叫感叹句。
(1)感叹句的句末通常用感叹号,但也可用句号,一般用降调。感叹句的构成方法有两种,如:
What a magnificent church it is! 这教堂真宏伟!
What wonderful square it is! 多漂亮的广场啊!
How lovely they are! 真好看!
How fast you are working! 你们干得真快!

(2)感叹代词 what 和它所修饰的名词短语置于句首,其他成分次序不变。句中某些成分经常被省略,如:
What a grand monument it is! 它是一座多么雄伟的纪念碑啊!
What a popular trend neoclassical architecture is now! 现在新古典主义风格是一个多么流行的趋势!
What a pity! 真可惜!

(3)感叹代词 how 和它所修饰的成分置于句首,其他成分次序不变。句中某些成分经常被省略,如:
How well she draws the building sheet! 她建筑图纸画得真好!

How peaceful the Beijing courtyard houses are! 北京的四合院真宁静啊!
Oh, Pantheon, how wonderful! 啊,万神殿,太棒了!

Practice

Turn these statements into the exclamatory sentences, beginning them with what or how.

1) The flowers look beautiful in the garden.
   _____

2) We have had an exciting day.
   _____

3) I was foolish to think like that.
   _____

4) It is lovely weather.
   _____

5) We are proud of our great motherland.
   _____

6) He has a fine voice.
   _____

Part 5 Applying

1. Enjoy These Sayings and Try to Learn Them by Heart.

1) *Be slow in choosing a friend, slower in changing.* ——Benjamin Franklin
   选择朋友要谨慎,更换朋友更要谨慎。——富兰克林

2) *Friendship is both a source of pleasure and a component of good health.*
   —— Ralph Waldo Emerson
   友谊既是快乐之源泉,又是健康之要素。——爱默生

3) *He that will not allow his friend to share the prize must not expect him to share the danger.* —— Aesop, Ancient
   不肯让朋友共享果实的人,不要指望朋友与他共患难。——伊索

2. Creative Writing — the Style of Medieval European Architecture

Directions: Please find out the representative structures of the medieval European architecture and neoclassical architecture and introduce one type of ancient Roman architecture about its time, specialties and style, etc. Write an essay about the architecture with its picture in English. Print it in Paper A4 and send a copy to the teacher's e-mail box.

Lesson 4
The American Architectural Style

- **Part 1 Listening and Speaking**

    Topic   The Pyramids and Temples

- **Part 2 Intensive Reading**

    Text A   The Development of the American Constructions

- **Part 3 Extensive Reading**

    Text B   Famous Architects

- **Part 4 Sentence Structures**

    Sentence Patterns

- **Part 5 Applying**

    Enjoy These Sayings and Try to Learn Them by Heart

    Creative Writing — Features of Museums in America

Lesson 4
The American Architectural Style

Part 1 Listening and Speaking

Topic The Pyramids and Temples

Directions: This part is to train your listening ability. The dialogue will be spoken twice. After the first reading, there will be a pause. During the pause, you must read after it and then listen to it again in order to put the sentences into Chinese.

A: Welcome home! How was your trip to South America? Did you get to see some amazing architecture?

B: It was excellent. And yes, I saw a great deal of architecture. We spent most of our time in Mayan territory.

A: I've heard you don't have to be there long to really get the flavor of the place.

B: The predominant structures there were the pyramids and temples. The stone work was absolutely superb.

A: The Mayan people must have been highly intellectual to build such structures.

B: What I found most interesting was the way they were able to make them appear rustic and yet exquisite at the same time. The pyramids were remarkably compact and predominantly made of stone.

A: I hope you take the time to reflect on your visit there and really savor the experience. It's not often that you get a chance to visit such an elaborate and ethnic site. I'm glad you went!

Chinese Versions

A: _____
B: _____
A: _____
B: _____
A: _____
B: _____
A: _____

Part 2 Intensive Reading

Text A

The Development of the American Constructions

The Architecture of the United States includes a wide variety of architectural styles over its history. Architecture in the U.S.A. is regionally diverse and has been shaped by many external forces, and can therefore be said to be eclectic, something unsurprising in such a multicultural society.

Exterior styles and related building forms and floor plans are in part a product of cultural tastes and values that reflect a particular place, time, and population. Styles are somewhat analogous to clothing fads, which come and go over time, and sometimes return. When the spread of cultural ideas and fashions all over the country was slower, certain architectural styles also remained in vogue for multiple decades or longer and often revealed a distinctly regional identity.

By the Victorian Era of the mid-to-late nineteenth century, multiple styles became simultaneously popular and readily available throughout the United States, ushering in what historians refer to as the "Eclectic Era" of architecture, when Americans tended to numerous modern or revival styles. This co-existing fascination with so-called "period styles" and early modernism continued unabated until the Great Depression. And after that, in fact relatively little building construction took place between 1929 and 1945.

Not until after World War II did America see another national building boom, by which time automobile suburbs; modern-era housing and office towers were the rule. America's modern era of functionalism and a general aversion to historic references dominated the built environment from the 1940s through the 1980s. The familiar "glass box" office tower and ubiquitous suburban ranch house are still powerful symbols of this anti-stylistic era when "form followed function".

Changes were happening by the 1970s, however, leading America to react against modern architecture. Historic styles became gradually popular once again, coinciding with the now-booming historic preservation movement. Colonial Revival elements adorned otherwise modern ranch houses, and by the 1990s a vague "Postmodern Era" was in full swing. Postmodern architecture is generally characterized by an unrelated and exaggerated use of historical styles, or imitated reproductions of older buildings. The current rise of postmodern historicism has coincided with a revived interest in traditional town building practices known as "nontraditional" development, or the New Urbanism. A return to conventional town building with its high-rise, mixed-used lofts and condos is now occurring, and hundreds of nontraditional neighborhoods are under construction or already completed, with designs that variously emphasize walking, mass transit, mixed uses, community livability, public space, and also hopeful affordability.

Architectural styles in America is still changing and developing. And we can't help wondering what will be America's next major cultural interest, and how will the built environment reflect that interest?

(435 words)

Images Connected with Internet 欣赏相关视频截图

Victorian Era   Colonial Revival   New Urbanism

Word Study

Words	Pronunciation	Versions
external	[eks'tə:nl] *adj.*	happening or arising or located outside or beyond some limits or especially surface 外界的,外来的,非本质的
eclectic	[ek'lektik] *adj.*	selecting what seems best of various styles or ideas 不拘一格的,兼收并蓄的,折衷的
analogous	[ə'næləgəs] *adj.*	~ (to/with sth) similar in someway to another thing or situation and therefore able to be compared with it 相似的,类似的
vogue	[vəug] *n.*	~ for sth, a fashion for sth. 时尚,流行,时髦
reveal	[ri'vi:l] *v.*	make known to the public information that was previously known only to a few people 展现,揭示,暴露,泄露
distinctly	[dis'tiŋkt] *adv.*	clear to the mind; with distinct mental discernment 清晰地,无疑地
identity	[ai'dentiti] *n.*	the distinct personality of an individual regarded as a persisting entity 个性,身份
simultaneously	[ˌsiməl'teinjəsli] *adv.*	occurring or operating at the same time 同时地

available	[ə'veiləbl] adj.	obtainable or accessible and ready for use or service; not busy; not otherwise committed 可用的, 可与之联系的
usher	['ʌʃə] v.	show (someone) to their seats, as in theaters or auditoriums 引领, 陪同, 迎接
unabate	[ˌʌnə'beitid] adj.	[not usually before noun](written)without becoming any less strong 不减, 未变弱
automobile	['ɔ:təməubi:l] n.	a motor vehicle with four wheels 汽车
aversion	[ə'və:ʃən] n.	a feeling of intense dislike 厌恶, 反感
dominate	['dɔmineit] v.	be larger in number, quantity, power, status or importance; be in control 在……中占主要地位, 支配
ubiquitous	[ju:'bikwitəs] adj.	[usually before noun] (formal or humorous) seeming to be everywhere or very common 似乎无处不在的, 十分普遍的
react	[ri'ækt] v.	show a response or a reaction to something; undergo a chemical reaction 反抗, 反应, 起反作用
vague	[veig] n.	not clearly understood or expressed; not precisely limited, determined, or distinguished (形状等)模糊不清的 (想法等)不明确的, 暧昧的, 含糊的
exaggerate	[ig'zædʒəreit] v.	to enlarge beyond bounds or the truth 夸张, 夸大其词
exterior	[iks'tiəriə] n. / adj.	the region that is outside of something 外部, 面, 外型 外部的, 表面的

Proper Nouns

1. Victorian Era [vik'tɔ:riən'iərə] 维多利亚时代（1837-1901）
2. Eclectic Era【建】折衷主义建筑时代
   十九世纪上半叶至二十世纪初, 在欧美一些国家流行的一种建筑风格。折衷主义建筑师任意模仿历史上各种建筑风格, 他们不讲求固定的法式, 只讲求比例均衡, 注重纯形式美。
3. the Great Depression 美国经济大萧条时期 (1930s)
4. World War II 第二次世界大战(1939—1945)
5. functionalism【建】['fʌŋkʃənəlizəm] 功能主义建筑
6. New Urbanism【建】新都市主义 (20世纪80年代兴起的城市规划运动)
7. Revival styles【建】复兴风格
8. Period styles【建】室内设计风格或流派

Useful Expressions

① analogous to 类似的，相似的
The heart is analogous to a pump. 心脏和水泵有相似之处。

② refer to 提到，涉及
Don't refer to this matter again, please. 请别再提这件事。

③ tend to 有……的倾向
Modern furniture design tends to simplicity. 现代家具设计越来越简单。

④ take place 发生；举行
When will the basketball game take place? 篮球赛何时举行？

⑤ react against 反抗
He reacted against the social bad habit by writing. 当时他用文章来反抗社会陋习。

⑥ coincide with 与……相一致，(两件或更多的事情)同时发生
My free time does not coincide with his. 我和他不是同时有空。

⑦ in full swing 正在全力进行中
The building project is in full swing. 这项建筑工程正在全力进行中。

Notes on Text A

① Styles are somewhat analogous to clothing fads, which come and go over time, and sometimes return.
   分析 somewhat "有点, 稍微"; which 引导非限制性定语从句, 修饰先行词 fads。

② When the spread of cultural ideas and fashions all over the country was slower, certain architectural styles also remained in vogue for multiple decades or longer, and often revealed a distinctly regional identity.
   分析 When 引导时间状语从句, certain architectural styles 为句子的主语, remained... revealed 为句子的并列谓语。

③ By the Victorian Era of the mid-to-late nineteenth century, multiple styles became simultaneously popular and readily available throughout the United States, ushering in what historians refer to as the "Eclectic Era" of architecture, when Americans tended to numerous modern or revival styles.
   分析 ushering... 为现在分词短语作状语, 表示伴随; what 引导介词宾语从句; when 引导时间状语从句。

④ Not until after World War II did America see another national building boom, by which time automobile suburbs, modern-era housing and office towers were the rule.
   分析 not until... 表示"直到……才", 它位于句首时, 后面句子要使用部分倒装; by which time 引导非限制性定语从句, 修饰先行词 boom。

⑤ Historic styles became gradually popular once again, coinciding with the now-booming historic preservation movement.

分析 coinciding with ... movement 现在分词短语,在句中作伴随状语;now-booming 合成名词,指 popular period,当今盛行,风靡期。

❻ A return to conventional town building with its high-rise, mixed-used lofts and condos is now occuring, and hundreds of nontraditional neighborhoods are under construction or already completed, with designs that variously emphasize walking, mass transit, mixed uses, community livability, public space, and also hopeful affordability.

分析 全句为并列复合句。并列句由第二个 and 连接;and 后面的分句中含有一个由 that 引导的限制性定语从句,修饰先行词 designs;第一和第三个 and 连接名词,构成名词短语,在句中分别作介词 with 和动词 emphasize 的宾语。

Part 3 Extensive Reading

Text B

Famous Architects

There are a lot of architects around the world throughout the history, who have made great contribution to the development of world architecture. The following is a list of famous architects—well known individuals with a large body of published work—both in America and in China.

Liang Sicheng (梁思成 1901-972) is one of the most famous architects in China. He was  the son of Liang Qichao, a well-known Chinese thinker in the late Qing Dynasty. Liang is the author of China's first modern history on Chinese architecture and founder of the Architecture Department of Northeast University in 1928 and Tsinghua University in 1946. He was the Chinese representative in the Design Board which designed the United Nations headquarters in New York. He is regarded to be the "Father of Modern Chinese Architecture". To cite Princeton University, which awarded him an honorary doctoral degree in 1947, he was "a creative architect who has also been a teacher of architectural history, a pioneer in historical research and exploration in Chinese architecture and planning, and a leader in the restoration and preservation of the priceless monuments of his country."

Ieoh Ming Pei (贝聿铭 1917-), commonly known by his initials I. M. Pei, is a Pritzker Prize-winning Chinese-born American architect, known as the last master of high modernist architecture. In 1974 he designed The East Building of The National Gallery of Art, a national art museum, located on the National Mall in Washington, D.C. Another masterpiece of him is The Bank of China Tower (abbreviated BOC Tower) in 1989, which is one of the most recognizable skyscrapers in

Central, Hong Kong. Designed by I. M. Pei, the building is 305 meters (1,000.7 feet) high with two masts reaching 367.4 meters (1,205.4 feet) high. It was the tallest building in Hong Kong and Asia from 1989 to 1992, and it was the first building outside the United States to break the 305 meters (1,000 feet) mark.

Philip Cortelyou Johnson (1906-2005) was an influential American architect. With his thick, round-framed glasses, Johnson was the most recognizable figure in American architecture for decades. Johnson's early influence as a practicing architect was his use of glass; his masterpiece was the Glass House (1949) he designed as his own residence in New Canaan, Connecticut, a profoundly influential work. In 1930, he founded the Department of Architecture and Design at the Museum of Modern Art in New York City and later (1978), as a trustee, he was awarded an American Institute of Architects Gold Medal and the first Pritzker Architecture Prize, in 1979.

Richard Meier (1934-) is an American architect known for his rationalist designs and the use of the color white, which has been used in many architectural landmark buildings throughout history, including cathedrals and the white-washed villages of the Mediterranean region, in Spain, southern Italy and Greece. Identified as one of The New York Five in 1972, his commission of the Getty Center in Los Angeles, California catapulted his popularity among the mainstream. In 1984, Meier was awarded the Pritzker Prize, and in 2008, he won the gold medal in architecture from the Academy of Arts and Letters.

(525words)

Images Connected with Internet 欣赏相关视频截图

Chengdu Sichuan China by Liang Sicheng

Suzhou Museum by Ieoh Ming Pei

Crystal Cathedral 1977 by Philip Cortelyou Johnson

Word Study

Words	Pronunciation	Versions
individual	[ˌindi'vidjuəl] adj./n.	being or characteristic of a single thing or person—a person considered separately rather than as part of a group 个人，个人的，个体的，单独的
representative	[ˌrepri'zentətiv] n.	a person or thing that represents another or others 典型，代表物，代表
pioneer	[ˌpaiə'niə] n.	a colonist, explorer, or settler of a new land, region, etc (as modifier); an innovator or developer of something new 先驱
masterpiece	['mɑ:stəpi:s] n.	an outstanding work, achievement, or performance （个人或团体的）最杰出的作品
influential	[ˌinflu'enʃəl] adj.	having or exerting influence 有影响的，有权势的
trustee	[trʌs'ti:] n.	a person to whom the legal title to property is entrusted to hold or use for another's benefit 财产、业务等的受托管理人
rationalist	['ræʃənəlist] n.	someone who emphasizes observable facts and excludes metaphysical speculation about origins or ultimate causes 理性主义者，唯理主义者
commission	[kə'miʃən] n.	a formal request to sb to design or make a piece of work such as a building or a painting(请某人作建筑设计或作一幅画等的)正式委托
catapult	['kætəpʌlt] v./n.	to push or throw sb/sth very hard so that it moves through the air very quickly 猛投（British English）a small stick in the shape of a Y with a thin rubber band fastened over the two ends, used by children to throw stones 弹弓
mainstream	['meinstri:m] n.	(the mainstream) [sing.] the ideas and opinions that are thought to be normal because they are shared by most people; the people whose ideas and opinions are most accepted 主流思想，主流群体

Proper Nouns

1. Liang Sicheng 梁思成(1901—1972)，中国著名建筑师

2. Liang Qichao 梁启超(1873—1929), 中国晚清著名思想家, 戊戌维新运动领袖之一
3. Qing Dynasty 清朝, 中国末代王朝
4. United Nations headquarters 联合国总部, 位于美国纽约市曼哈顿区。
5. Princeton University ['prinstən] 普林斯顿大学, 位于美国新泽西州的普林斯顿, 是英属北美洲部分成立的第四所大学。
6. Ieoh Ming Pei 贝聿铭(1917—), 著名华裔美国建筑师
7. Pritzker Prize 普利兹克奖是每年一次颁给建筑师个人的奖项, 有建筑界的诺贝尔奖之称。
8. The National Gallery of Art (美国)国家美术馆, 位于美国国会大厦西阶, 国家大草坪北边和宾夕法尼亚大街夹角地带, 是世界上建筑最精美、藏品最丰富的美术馆之一。
9. The Bank of China Tower 香港中国银行大厦, 由贝聿铭建筑师事务所设计, 1990年完工。
10. Philip Cortelyou Johnson 菲力普·科特柳·约翰逊(1906—2005), 著名美国建筑师
11. New Canaan, Connecticut [nju'kenənkə'netikət] 康涅狄格州新迦南市
12. American Institute of Architects 美国建筑师学会, 为美国建筑界权威性的组织, 总部位于华盛顿哥伦比亚特区。
13. Richard Meier 理查德·迈耶(1934—), 著名美国建筑师
14. The New York Five 纽约五人派, 纽约五人组, 又称"白派", 是现代派原则的拥护者。代表人物有理查·迈耶、迈克·格雷夫斯、查理士·盖斯密、彼得·艾森曼及约翰·黑达克。
15. Getty Center 洛杉矶盖蒂中心, 它包括一座非常现代化的美术博物馆, 一个艺术研究中心和一所漂亮的花园。
16. Los Angeles [lɔs'ændʒələs] 洛杉矶, 位于美国加州西南部, 是美国新泽西州中部一个享有自治权的城市。

Useful Expressions

❶ make contribution to... 为……做出贡献

China will continue to make contribution to the promotion of the world peace. 中国将会继续为促进世界和平而做出贡献。

❶ be regarded as... 被认为是……

The old man does not want to be regarded as a burden. 这位老人不想成为别人的负担。

Notes on Text B

❶ He was the son of Liang Qichao, a well-known Chinese thinker in the late Qing Dynasty.

分析 Liang Qichao 和 a well-known Chinese thinker in the late Qing Dynasty 之间是同位语关系，well-known 为"著名的"，late Qing Dynasty 为"清朝末期"。

❷ Ieoh Ming Pei (贝聿铭) (1917-), commonly known by his initials I. M. Pei, is a Pritzker Prize-winning Chinese-born American architect, known as the last master of high modernist architecture.

分析 句中两个 known 都是过去分词做后置定语，Pritzker Prize-winning 中的 winning 现在分词做定语，Chinese-born 中的 born 是过去分词做定语。

❸ Richard Meier (1934-) is an American architect known for his rationalist designs and the use of the color white, which has been used in many architectural landmark buildings throughout history, including cathedrals and the white-washed villages of the Mediterranean region, in Spain, southern Italy and Greece.

分析 which 引导非限制性定语从句，including 为现在分词做伴随状语，cathedrals 表示"大教堂"，Mediterranean region 指的是地中海地区。

Part 4 Sentence Structures

Sentence Patterns 句型的类型

句子是一个语法单位，也是人们进行交际和表达思想的基本语言单位，它由词按语法规律构成，能单独存在并能表达一个相对完整的意思，是一定的语法结构、语言结构和语义结构的统一体。依据其结构，句子可以分为三类：简单句(simple sentences)、并列句(compound sentences)和复合句(complex sentences)。

1. Simple Sentence 简单句

只有一个主谓结构，且句子各成分均由单词或短语充当的句子称作简单句。

❖ 简单句的结构是：主语（或并列主语）+谓语（或并列谓语），如：

It took three years to construct this highway. 修这条公路花了三年时间。

Many well-known building companies are multinationals. 很多知名建筑公司都是跨国公司。

Time and tide wait for no man. 岁月不等人。（并列主语）

He went to town and mailed a letter to his family this morning. 今天上午他进城给家里寄了一封信。（并列谓语）

❖ 某些简单句结构特殊，只包含一个词或一个成分，如：

Hello! 喂！

Hands up! 举起手来！

So long! 再见！

2. Compound Sentence 并列句

并列句是由并列连词连接的两个或两个以上的简单句组成。

❖ **并列句的结构是：简单句+并列连词+简单句。**

并列句的各个分句(即简单句)在形式上是平等关系。但是，有种种不同的平等关系，因而要用不同的并列连词来表示，也可以用标点符号(如逗号、分号等)来表示这种并列关系，如：

He majors in architecture and his twin brother majors in architectural engineering. 他主修建筑学，他的孪生弟弟主修建筑工程学。

You may go with us to the Great Wall, or you may stay at the hotel today. 今天你可以和我们去长城，也可以待在宾馆里。

It was getting late; we had to get to the Construction Ministry before five. 天色已晚，我们必须五点之前赶到建设部。

❖ **并列连词是构成并列句的一个关键**

英语中的并列句绝大多数是通过并列连词来实现的，因而并列连词在并列句的构成中起着相当重要的作用，是构成并列句的关键。以下按类别将简单介绍某些并列连词的用法。

(1) 联合并列句

主要由 and, not only... but also... (不但……而且……), when (=and just at this time 就在这时)等连词连接，如：

The Great Hall of the People is one of the most magnificent buildings in Beijing, and it attracts thousands of visitors every year.

人民大会堂是首都最宏伟的建筑之一，每年吸引着成千上万的游客。

Not only did he go to Tian'anmen Square, but he went to Summer Palace last National holiday.

去年国庆节他不仅去了天安门广场，还去了颐和园。

(2) 转折并列句

主要由 and (和，与，同，又), but (但是，而且), yet (然而，可是)和 while (而，却) 等连词连接，如：

I would have written to my teacher of architecture before but I have been ill. 我本该早写信给建筑老师的，但我生病了。

I have failed in the design, yet I shall try again. 我的设计失败了，但我还要尝试。

(3) 选择并列句

主要由 or (或者，还是，否则), either...or... (不是……就是……), neither...nor... (既不……也不……), otherwise (要不然)等连词连接，如：

Seize the chance to visit the building of Tang Dynasty in Xi'an, otherwise you will regret it. 要抓住这次参观西安唐代建筑的机会，否则你会后悔的。

Neither does he like this building material nor does he like this architectural ornament. 他既不喜欢这种建筑材料也不喜欢这一建筑装饰。

(4) 因果并列句

主要由 for（因为）, so（因此）和 therefore（因此）等连词连接，如：

He shook his head, for he thought differently toward the architectural design. 他摇了摇头，因为他对这一建筑设计有不同想法。

He told me to read some books on building law, so I borrowed some from the library. 他让我阅读一些有关建筑法规方面的书籍，于是我从图书馆借了一些。

❖ 并列连词中还有一类介乎于并列连词与从属连词或介词之间的词组，如：as well as, rather than, more than, as much as 等，它们也可以作为并列连词使用。

3. Complex Sentence 复合句

在复合句中，主句是全句的主体；从句从属于主句，是全句的一个成分，不能独立。从句通常用关联词来引导，并由关联词将从句和主句连在一起。从句的结构通常是：关联词+主语+谓语。

连接主句和从句的关联词有以下几类：

(1) 从属连词，如：

after, as, as long as, as soon as, although, because, before, if, since, though, that, whether, when, while 等。

(2) 连接代词，如：

who, whom, whose, which, what 等

(3) 连接副词，如：

when, where, why, how 等

(4) 关系代词，如：

who, whom, whose, which, that 等

(5) 关系副词，如：

when, where, why 等

从句在复合句中可以充当主语，宾语，表语，同位语，定语，状语等。因此可分为：主语从句，宾语从句，表语从句，同位语从句，定语从句和状语从句。

❖ 名词性从句

主语从句、宾语从句、表语从句和同位语从句在复合句中的功能与名词相同，因此，它们可以称为名词性从句。从句和主句之间不用逗号隔开，如：

Whether he will be able to come remains a question. 他是否能来，还是个问题。

It is a pity that you missed such a good chance. 很遗憾，你错过了这样一个好机会。

Which room I will hire is not decided. 我要租哪间房子还未决定。

The question is who will support us. 问题是谁会来支持我们。

You know how clever some animals are. 你知道有些动物有多聪明。

I'll take back what I said. 我收回我说的话。

You may think it strange that anyone would live there. 你也许感到奇怪，居然有人愿意住在那里。

I have no idea who he is. 我不知道他是谁。

We all know the truth that the earth goes round the sun. 我们都知道地球绕着太阳转这一真理。

❖ 形容词性从句

定语从句一般放在所修饰的名词或代词后面，所修饰的名词或代词叫先行词。定语从句在复合句中的功能和形容词相同，因此，定语从句可以称为形容词性从句。

定语从句分限制性定语从句和非限制性定语从句两种。限制性定语从句是先行词不可缺少的定语，它和主句的关系十分密切，不可用逗号与主句隔开。其关系代词在从句中充当主语时可以省略，如：

This is the best novel (that) I have ever read. 这是我看过的小说中最好的一部。

The Great Wall has many gate ways which connect the main roads of North China. 长城有许多关口，这些关口与华北的主要道路相连。

It is actually connected with Asia at the spot where the Suez Canal was dug. 这实际上是在开凿苏伊士运河的地方与亚洲相连。

We can not accept the excuse why you have changed your plan. 我们无法接受你改变计划的借口。

非限制性定语从句是先行词（或者短语、句子等）的附加说明，如果省去，主句的意思仍然会很清楚，它和主句之间通常用逗号隔开，关系词不能省略，通常不用关系词that或why引导，如：

She is recovering quickly, which I am glad to hear. 她康复得很快，我听了很高兴。

There is a beautiful lake at the foot of the hill, whose depth (=the depth of which) has never been measured. 山脚下有一个美丽的湖，其深度从未测量过。

He is an American, which (as) I know from his accent. 我从他的口音得知他是个美国人。

He tore up my book, which upset me. 他把我的书撕破了，这使我很不高兴。

❖ 副词性从句

状语从句在复合句中的功能和副词相同，因此，状语从句可以称为副词性从句。它修饰主句中的动词、形容词、副词或者主句全句。状语从句可分为：时间状语从句，地点状语从句，原因状语从句，目的状语从句，结果状语从句，条件状语从句，方式（或伴随）状语从句，让步状语从句和比较状语从句，如：

While ants grow, they change their forms three times. 蚂蚁成长时形状要改变三次。

Where there is plenty of sun and rain, the fields are green. 哪里有充足的阳光和水，哪里的田地就绿油油的。

Since (Now that) all the members have arrived, let's have our meeting. 既然人都到齐了，咱们开会吧。

We should sit near the front in order that we can catch the speaker better. 我们要靠前坐，以便能听得更清楚些。

It looks as if it is going to rain. 看起来好像要下雨。

We are determined to fulfill the task, whatever (no matter what) happens. 不管发生什么事，

我们都决心完成这项任务。

The project was completed earlier than we had expected. 这项工程提前完工了。

The amount of cement is three times more than that of sand. 水泥的量是沙子的三倍。

Practice

Choose the best answer to complete each sentence.

1. Would you like to have a break _____ shall we go on with the work?
   A. and   B. then   C. therefore   D. or

2. Five minutes earlier, _____ we could have caught the last train.
   A. or   B. but   C. and   D. so

3. _____ many times, but he still couldn't understand it.
   A. Having been told   B. Though he had been told
   C. He was told   D. Having told

4. Tom's mother kept telling him that he should work harder, but _____ didn't help.
   A. he   B. which   C. she   D. it

5. _____, so Mary was very sad.
   A. Her mother was badly ill   B. Her mother being ill
   C. Her mother's being ill   D. Because her mother was ill

6. Sugar _____ milk.-Only milk, please, _____ I used to like sugar.
   A. and; and   B. and; but   C. or; and   D. or; but

7. They seldom have lunch at home, _____?
   A. haven't they B. don't they   C. do they   D. have they

8. _____ knows the fact will surely tell you.
   A. That who   B. Who that   C. That   D. Whoever

9. My Aunt Jennifer looks _____ she were only 20 years old although she is 40 now.
   A. that   B. as if   C. as   D. like

10. There are a lot of reasons _____ can make me cheer up.
    A. why   B. /   C. for which   D. that

Part 5 Applying

1. Enjoy These Sayings and Try to Learn Them by Heart.

Living without an aim is like sailing without a compass.   —— *J. Ruskin*
生活而无目标,犹如航海之无指南针。   ——鲁斯金

> *Victory won't come to me unless I go to it.* —— *Moore*
> 胜利是不会向我走来的,我必须自己走向胜利。 ——穆尔

> *Life is just a series of trying to make up your mind.* —— *T. Fuller*
> 生活只是由一系列下决心的努力所构成。 ——富勒

2. Creative Writing — Features of Museums in America

Directions: Find out one of the representative modern buildings in America, such as a museum, a bank and so on. Write an essay about the architecture in English and introduce its time, specialties and style, etc. Print it in Paper A4 and send a copy to the teacher's e-mail box.

3. How much do you know?

Architects	Well-known books	Representatives	Typical Designing Style
Liang Sicheng (1901-1972) 梁思成			
Ieoh Ming Pei (1917-) 贝聿铭			
Philip C Johnson (1906-2005) 菲力普·约翰逊			
Richard Meier (1934-) 理查德·迈耶			

Lesson 5
The Architectural Art in Asia

- **Part 1 Listening and Speaking**
  Topic   Our New Project

- **Part 2 Intensive Reading**
  Text A   Potala Palace and Forbidden City

- **Part 3 Extensive Reading**
  Text B   Buildings in Ancient Kyoto, Japan

- **Part 4 Sentence Structures**
  Noun Clauses

- **Part 5 Applying**
  Enjoy the Idioms and Try to Use Them
  Creative Writing — Features of Temples and Shrines in Japan

Lesson 5
The Architectural Art in Asia

Part 1 Listening and Speaking

Topic Our New Project

Directions: This part is to train your listening ability. The dialogue will be spoken twice. After the first reading, there will be a pause. During the pause, you must read after it and then listen to it again in order to put the sentences into Chinese.

A: Our next project will be to build a set for a Shakespearean play. They want the background to look very medieval. Do you have any suggestions?

B: Well, let's see what we can do to imitate the architecture of the time. I know they used lots of decorative arches.

A: One of the scenes takes place in a cathedral and I get the impression that they want us to submit some illustrations before we begin constructing the set.

B: That was precisely what I was going to say. I think we should plan for the aisle to be flanked with buttresses. That way everyone in the cathedral will be able to see without interruption.

A: It sounds good to me. I'll start work on the illustrations while you begin gathering materials. Let's plan on meeting again at this same time next week.

B: OK. See you then.

Chinese Versions

A: _____
B: _____
A: _____
B: _____
A: _____
B: _____

Part 2 Intensive Reading

Text A

Potala Palace and Forbidden City

In 641 AD, after marrying Princess Wencheng, Songtsen Gampo decided to build a grand palace to accommodate her and let his descendants remember the event. The monastery-like palace, reclining against and capping Red Hill, was the religious and political center of old Tibet and the winter palace of Dalai Lamas. The palace is more than 117 meters (384 feet) in height and 360 meters (1180 feet) in width, occupying a building space of 90 thousand square meters. Potala is composed of White Palace and Red Palace. The former is for secular use, while the latter is for religious.

Potala Palace was built near mountains. It includes overlapping group of buildings and towering temples. The Potala Palace has solid pier, thick granite walls and resplendent top, with a strong decorative effect of the huge gold Aquarius. The sharp contrast of three colors—red, white and yellow is very attractive. Division of jointly building, construction type of sockets on layer body all reflect the characteristics of charming old buildings of Tibet.

The design and construction of the palace are in accordance with the law of sunlight at high altitude. The wide and solid base wall, under which there are tunnels and vents extending in all directions. Inside, there are columns, brackets, beams, rafters, which form the support frame. The hard soil is used to build floor and roof. At the top of every hall and bedroom is window, which is easy to get light and fresh air. There are a variety of sculptures on columns and beams. Color murals on wall take up more than 2,500 square meters.

Lying at the center of Beijing, the Forbidden City, in Chinese, was the imperial palace during the Ming and Qing Dynasties. Rectangular in shape, it is the world's largest palace complex and covers 74 hectares. Surrounded by a six-meter-deep moat and a ten-meter-high wall are 9,999 buildings.

Construction of the palace complex began in 1407, the 5th year of the Yongle reign of the third emperor of the Ming Dynasty. It was completed fourteen years later in 1420. It was said that a million workers including one hundred thousand artists were driven into the long-term hard labor. Stone needed was quarried from Fangshan, a suburb of Beijing. It was said a well was dug every fifty meters along the road in order to pour water onto the road in winter to slide huge stones on ice into the city. Huge amounts of timber and other materials were freighted from faraway provinces. Ancient Chinese people displayed their very considerable skills in building the Forbidden City. Take the grand red city wall for example. It has an 8.6-meter-wide base reducing to 6.66 meters wide at the top. The angular shape of the wall totally frustrates attempts to climb it. The bricks were made from white lime and glutinous rice while the cement is made from glutinous rice and egg whites. These incredible materials make the wall extraordinarily strong.

Since yellow is the symbol of the royal family, it is the dominant color in the Forbidden City. Roofs are built with yellow glazed tiles; decorations in the palace are painted yellow; even the bricks on the ground are made yellow by a special process.

(539 Words)

Word Study

Words	Pronunciation	Versions
fundamentally	[ˌfʌndə'mentəl] adv.	in a way that affects the basis or essentials; utterly 基本上，根本上
respective	[ris'pektiv] adj.	belonging or relating separately to each of several people or things 各自的，分别的
descendant	[di'send(ə)nt] n.	a person, animal, or plant when described as descended from an individual, race, species, etc. 后裔，子孙
monastery	[mə'nəsteri] n.	a building in which monks (members of a male religious community) live together 隐修院，修道院，寺院
secular	['sekjulə] adj.	not concerned with or related to religion 世俗的，非宗教的
overlap	['əuvə'læp] v.	(of two things) to extend or lie partly over (each other); to cover and extend beyond (something); to coincide partly in time, subject, etc. 与……部分相一致[巧合]
pier	[piə(r)] n.	a long low structure built in a lake, river or the sea and joined to the land at one end; a large strong piece of wood 码头，桥墩，墙墩，柱子
granite	['grænit] n.	a light-colored coarse-grained acid plutonic igneous rock consisting of quartz, feldspars 花岗岩，花岗岩灰色
resplendent	[ris'plendənt] adj.	having a brilliant or splendid appearance 辉煌的，灿烂的
altitude	['æltitju:d] n.	the height above sea level 海拔，海拔高度 a place that is high above sea level 高地
tunnel	['tʌnl] n.	an underground passageway, esp one for trains or cars that passes under a mountain, river 烟囱，风洞，管道，隧道
bracket	['brækit] n.	a support projecting from the side of a wall or other structure 支柱，托架
rafter	['rɑ:ftə] n.	any one of a set of sloping beams that form the framework of a roof 椽
mural	['mjuərəl] n.	a large painting or picture on a wall 壁画
rectangular	[rek'tæŋgjulə] adj.	having four right angles; a rectangular figure twice as long as it is wide 矩形的，长方形的
timber	['timbə] n.	wood, esp. when regarded as a construction material 木料，木材

angular	['æŋgjulə] *adj.*	lean or bony; having an angle or angles 尖的, 倾斜的, 多角的
frustrate	[frʌs'treit] *v.*	to hinder or prevent (the efforts, plans, or desires) of; thwart; to upset, agitate, or tire 破坏, 挫败, 使落空
cement	[si'ment] *n.*	a grey powder made from lime and clay that becomes hard when it is mixed with water and allowed to dry, and that is used in building 水泥
lime	[laim] *n. / v.*	a white substance obtained by heating limestone 石灰 to add the substance lime to soil 掺加石灰
glutinous	['glu:tinəs] *adj.*	resembling glue in texture; sticky 粘的, 粘质的
incredible	[in'kredəbl] *adj.*	beyond belief or understanding, unbelievable, marvelous, amazing 难以置信的, 不可思议的, 惊人的
dominant	['dɔminənt] *adj.*	having primary control, authority or influence; governing; ruling; predominant or primary 支配的, 有权威的
contrast	['kɔntrʌst] *n.*	a difference between two or more people or things in order 对比, 对照
	[kən'trɑ:st] *v.*	to show the differences between them 对比, 对照

Proper Nouns

1. Potala Palace [pəu'dalə'pæləs] 布达拉宫
   The Potala Palace is located in Lhasa, Tibet Autonomous Region. It was named after Mount Potala, the abode of Chenresig or Avalokitesvara... (Lhasa: the sacred city of Lamaism)

2. Forbidden City [fə'bidn'siti] 紫禁城
   a walled section of Beijing that encloses the palace that was formerly the residence of the emperor of China

3. Kyoto ['kjəutəu] 古京都
   the former capital of Japan; now a prefecture on Honshū, Japan; the city and Kyoto prefecture, Japan.

4. Princess Wencheng ['prinses 'wentʃən] 文成公主
   公元七世纪初，吐蕃时期松赞干布迎娶了唐文成公主和尼泊尔赤尊公主，她们分别将释迦牟尼佛像和觉阿佛像带入了西藏，佛教开始在这块土地上传播。

5. Songtsen Gampo ['sʌŋtzən 'gæm'pu] 松赞干布

   Songtsen Gampo (Tibetan, Wylie: Srong-btsan-sgam-po, 605 or 617 - 649) was the founder of the Tibetan Empire (Tibetan: Bod) 吐蕃

6. Dalai Lamas ['dale] 达赖喇嘛

   The Dalai Lama is a lineage of religious leaders of the Gelug school of Tibetan Buddhism. The Dalai Lamas were also the political leaders of Lhasa...

7. Aquarius [ə'kweəriəs] 宝瓶宫, 宝瓶(星)座

   the 11th sign of the zodiac, the Water Carrier 黄道第十一宫

Useful Expressions

❶ be based on 根据, 以……为基础

Judgment should be based on facts, not on hearsay. 判断应该以事实为依据, 而不应该依靠道听途说。

❷ generally speaking 一般来说

Generally speaking, I think you're right. 一般来说, 我认为你是对的。

❸ foundationally speaking 从根本上来说

Foundationally speaking, the project has been cancelled because of lack of funds. 该项目已经被取消的根本原因是缺少资金。

❹ focus on 集中

Focus your attention on your work. 把注意力集中在你的工作上。

❺ be composed of 由……组成

The Election Committee shall be composed of 800 members. 选举委员会将由800人组成。

❻ in accordance with 与……一致, 依照

He acted in accordance with his beliefs. 他按照自己的信念行事。

❼ a variety of 各种各样的, 种种

There is(are) a variety of flowers in the garden. 花园里有各种各样的花卉。

Notes on Text A

❶ In 641 AD, after marrying Princess Wencheng, Songtsen Gampo decided to build a grand palace to accommodate her and let his descendants remember the event.

   分析 marry sb. 娶或嫁给某人; decide to do sth. 决定做某事; grand 豪华的, 雄伟的; event 重大事件

❷ The former is for secular use while the latter is for religious.

分析 the former..., the latter... 前者……,后者……

❸ The design and construction of the palace are in accordance with the law of sunlight at high altitude.

分析 altitude 高度,海拔;law 在这里的意思是"规律"。

❹ Surrounded by a six meter deep moat and a ten-meter high wall are 9,999 buildings.

分析 此句是完全倒装句,主语是 buildings,谓语动词是 are surrounded;moat 护城河;ten-meter high 十米高

❺ It was said that a million workers including one hundred thousand artists were driven into the long-term hard labor.

分析 It is/was said that... 据说;drive 意思是 force sb. to do sth. 迫使/强迫某人做某事;long-term 长期的

❻ It was said a well was dug every fifty meters along the road in order to pour water onto the road in winter to slide huge stones on ice into the city.

分析 well 名词,意思是"井";in order to 表示"为了……的目的"

❼ The bricks were made from white lime and glutinous rice while the cement is made from glutinous rice and egg whites.

分析 be made of/from 由……制成,看出构成的原料用介词 of,看不出构成的原料用介词 from;white lime 白石灰;glutinous rice 糯米,黏米;egg whites 鸡蛋清

❽ Red Hill 红山,位于新疆维吾尔自治区乌鲁木齐市的一座褶皱断层山,通体由二叠纪的紫红色砂砾岩构成,西端断崖出现赭红色,故名"红山"。

Part 3 Extensive Reading

Text B

Buildings in Ancient Kyoto, Japan

Kyoto, the ancient ruins, is located in the former regional capital of Heian-kyo (see the right-upper figure). It was established in 794 AD, and since the time until the Edo period (1600-1868), it has been the capital of Japan. Ancient Kyoto was built following the example of the ancient capital of China. The initial design of Ancient Kyoto is based on Chang'an and Luoyang in the Chinese Sui and Tang times. The entire building was rectangular with a view to the north-south road axis Suzaku. Road, which divided the ancient Kyoto into two parts—east city and west city. East city followed the example of Luoyang, and west city Chang'an. The

Imperial Palace stands in the middle of east and west city. City streets were Chessboard-shaped and they are in order either from east to west or from north and south. And there is a clear division of the Imperial Palace, officials, residential and commercial areas.

Kyoto Gosho is Japan's old imperial palace, also known as the National Palace Museum. It has been the residence of successive emperors, and later became the emperor's palace. Kyoto Palace was burned 7 times before and after. The present Imperial Palace was reconstructed by Emperor Takaaki, with an area of 110,000 square meters, surrounded by walls.

Heian Shrine is a magnificent building, as the representative of garden in the Meiji era. Pavilion on the Lake Palace, are modeled on the structure of the temple in Xi'an, China.

Nijo Castle, (see the right-down figure: a part of the Nijo) built in 1603, is magnificent and just as simple in sharp contrast to the National Palace Museum. Huge rocks (see the left-down figure: a part of the Castle Wall) are used to build the castle wall, around which is the moat, 500 meters long from east to west, 300 meters long from north to south. There is the Tang-styled building on the river.

Referred to as "three steps, a temple; seven steps, a shrine", the Kyoto has more than 1,500 Buddhist temples, shrines over 2,000 blocks. This is the cradle of Japanese culture and the arts, the center of Buddhism.

(332 Words)

Images Connected with Internet 欣赏相关视频截图

Word Study

Words	Pronunciation	Versions
ruin	['ruːin] v./n.	to damage something so badly that it loses all its value, pleasure, etc 毁坏,破坏,糟踏,残垣断壁,废墟
establish	[is'tæbliʃ] v.	to make secure or permanent in a certain place, condition, job, etc. 建立,创立
axis	['æksis] n.	a straight line through a body or figure that satisfies certain conditions 轴,轴线,中心线,中枢
residential	[rezi'denʃəl] adj.	suitable for or allocated for residence; relating to or having residence 住宅的,适合居住的
commercial	[kə'məːʃəl] adj.	of, connected with, or engaged in commerce; mercantile; having profit as the main aim 商业化的
successive	[sək'sesiv] adj.	following another without interruption; of or involving succession 接连的,相继的,逐次[步,位]的
magnificent	[mæg'nifisnt] adj.	splendid or impressive in appearance;superb or very fine; deserving praise 宏伟的,华丽的,值得赞扬的
cradle	['kreidl] n.	a baby's bed with enclosed sides or a place where something originates or is nurtured during its early life 摇篮,发源地

Proper Nouns

1. Heian-kyo 平安京
2. Edo period (1600-1868) 江户时代
3. Suzaku Road 朱雀路
4. Kyoto Gosho 京都御所
5. Emperor Takaaki 孝明天皇
6. Heian Shrine 平安神宫
7. Meiji era 明治时代
8. Nijo Castle 二条城
9. Buddhism 佛教

Useful Expressions

❶ in contrast to 和……形成对比[对照]

His white hair was in sharp contrast to his dank skin. 他的白发同他的黑皮肤形成鲜明的对比。

❷ by contrast 相比之下

You say by contrast, or in contrast to something, to show that you are mentioning a very different situation from the one you have just mentioned. 相比之下，你现在描述的情景和你刚才所说的完全不一样。

Notes on Text B

❶ The entire building was rectangular with a view to the north-south road axis Suzaku Road, which divided the ancient Kyoto into two parts—east city and west city.

分析 rectangular 矩形；which 引导非限制性定语从句，修饰先行词 axis；Suzaku Road 是 axis 的同位语。

❷ East city followed the example of Luoyang, and west city Chang'an.

分析 这是一个省略句型。..., and west city 后面省略了 followed the example of，省略的目的是避免重复。

❸ Huge rocks are used to build the castle wall, around which is the moat, 500 meters long from east to west, 300 meters long from north to south.

分析 be used to do 被用做；around which...to south 是由"介词+关系代词"引导的非限制性定语从句；500 meters long from east to west, 300 meters long from north to south 是名词短语，作 moat 的同位语，起补充说明作用。

❹ Referred to as "three steps, a temple; seven steps, a shrine", the Kyoto has more than 1,500 Buddhist temples, shrines over 2,000 blocks.

分析 Referred to as... 是过去分词短语，在句中作状语，修饰全句。

Part 4 Sentence Structures

Noun Clauses 名词性从句

在句子中起名词作用的从句叫名词性从句，包括主语从句、宾语从句、表语从句和同位语从句。名词性从句的词序是陈述句的词序。名词性从句的谓语动词形式受主句谓语动词形式的制约。名词性从句的连词大体有三类：

(1) that (无词义，在从句中不充当成分)

(2) whether/if (有词义，在从句中不充当成分)

(3) wh-words(有词义,在从句中充当成分),如:who, whom, whose, which, what, when, where, why, how等。

1. Subject Clause 主语从句

在句子中充当主句的从句称为主语从句。它的位置与陈述句基本结构中的主语相同。引导主语从句的关联词有以下几种:

❖ **由 that 引导主语从句,无词义,如:**

That the earth goes round the sun is well known. 众所周知,地球绕着太阳转。

主语从句置于句首,因句子过长而导致头重脚轻的缘故,常常用it作形式主语来引导,主语从句后置,如:

It is known that concrete is widely used as basic building material. 众所周知,混凝土是不可或缺的建筑材料。

It happened that one of my friends knew some customs in Japan. 碰巧我朋友了解一点日本风俗。

It was obvious that Ancient Chinese people displayed their very considerable skills in building the Forbidden City. 显而易见,古代中国人在建造紫禁城过程中显示出高超的技艺。

❖ **由 whether 引导的主语从句,if 一般不能引导主语从句,如:**

Whether or not she will go is up to her to decide. 她去还是不去由她自己决定。

这种从句也可以用it作形式主语,如:

It is uncertain whether the game will be held on schedule. 比赛是否能如期举行,这还不确定。

不能说:If it is true remains a question. 这是不是真的,仍是个问题。

但如用it作形式主语时,则后面可以用if引导的主语从句,如:

It remains a question if/whether it is true. 这是不是真的,仍是个问题。

❖ **由 who, whom, whose, which, what, whoever, whatever 等引导的主语从句,如:**

Who will go makes no difference. 谁去都没有关系。

What attitude you will take means a lot. 你采取什么态度意义重大。

Whoever breaks the rule should be punished. 无论谁违犯这条规定都要受到惩罚。

❖ **由 when, where, why, how 等引导的主语从句,如:**

When he'll be back depends much on the weather. 他什么时候能回来这要看天气情况。

How the huge stones were moved is still a mystery. 当时这些巨大的石头是怎么搬运的到现在仍然是个谜。

2. Object Clause 宾语从句

在句中充当宾语的从句称为宾语从句。它的位置与陈述句基本结构中的宾语相同。引导宾语从句的关联词和引导主语从句的相同。

- 由 **that** 引导的宾语从句，在口语中 **that** 常常省略，如：

He decided (that) he was going to be an architect after his graduation from college. 她决定大学毕业后当一名建筑师。

He explained that he hadn't yet won a scholarship. 他解释道他还没拿到学位。

- 由 **whether** 或 **if** 引导的宾语从句，如：

Bob asked John whether/if he had visited the Summer Palace when he was in China. 鲍勃问约翰在中国的时候去游览过颐和园吗？

I went to town last weekend and found if there was any ranch house to be sold. 上周末我进城去看看有没有出售平房的。

- 由 **who, whom, whose, which, what, whoever, whatever** 等引导的宾语从句，如：

Tom didn't know whose side he should take. 汤姆不知道自己该站在哪一边。

You may choose whichever you want. 你想要哪个就挑哪个。

She was shocked by what she had seen. 看到眼前发生的事情她非常震惊。

- 由 **when, where, why, how** 等引导的宾语从句，如：

Can you tell me where the booking office is? 请问，售票处在哪儿？

He asked when we would be in London. 他问我们什么时候能到达伦敦。

3. Predictive Clause 表语从句

在从句中充当表语的从句称为表语从句。它位于句子中的系动词后面，引导表语从句的关联词除和引导主语从句的相同外，还有 as if, because 等。

- 由 **that** 引导的表语从句，在口语中 **that** 常常省略，如：

The fact is that bridges are great symbols of mankind's conquest of nature. 事实是桥梁是人类征服大自然的象征。

The disadvantage of steel is (that) it oxidizes (氧化) easily. 钢的缺点是它易氧化。

- 由 **whether** 和 **as if** 引导的表语从句，如：

The question is whether we have plenty of water in the following days. 问题是我们的水还能维系几天。

It seems as if you have something to tell us. 你好像有话要对大家说。

- 由 **because** 引导的表语从句，如：

This is because he told a lie, so we don't trust him. 因为他撒谎，我们不信任他。

- 由 **who, whom, whose, which, what** 等引导的表语从句，如：

The question is which hearing aid is perfectly better to fit him. 问题是哪个助听器他戴着更适合。

The trouble is who can draw 62 pages of illustrations for the book. 问题是谁能为这本书画62幅插图。

- 由 **when, where, why, how** 等引导的表语从句,如:

  What I want to say is when we will start the project. 我想说的是我们什么时候开工。

  This is where Salt Lake City now lies. 这就是盐湖城现在所处的位置。

4. Apposition Clause 同位语从句

在句子中充当同位语的从句称为同位语从句。它一般位于名词后面,对该名词加以补充和说明。常见的名词有:announcement, belief, doubt, fact, message, idea, conclusion, thought, news, order, possibility, proof, promise, report, rumor, situation, word 等。that 在从句中不做成分,但是不能省略。

- 由 **that** 和 **whether** 引导的同位语从句,如:

  The situation that she had no word about the matter annoyed all of us. 对此事她一言不发,这让我们很恼火。

  We all know the truth that freeways generally are either depressed below ground level or elevated in the built-up parts of cities. 我们都知道在城市建筑密集区,高速公路一般建在地下或者高架起来。

  He must answer the question whether he agrees to it or not. 他必须回答他是否同意这样一个问题。

- 由连接代(副)词引导的同位语从句

  who, whom, whose, which, what, when, where, why, how 等都可以引导同位语从句,如:

  I have no idea who he is. 我不知道他是谁。

  Then rose a question how we could complete the project smoothly. 于是就产生了一个问题,我们如何能顺利完工。

Practice

Choose the best answer to complete each sentence.

1. A computer can only do _____ you have instructed it to do.
   A. how   B. after   C. what   D. when

2. —I drove to Zhuhai for the air show last week.
   —Is that _____ you had a few days off?
   A. why   B. when   C. what   D. where

3. I hate _____ when peope talk with their mouths full.
   A. it   B. that   C. these   D. them

4. ___ is a fact that English is being accepted as an international language.
   A. There   B. This   C. That   D. It

5. I still remember _____ this used to be a quiet village.
   A. when   B. how   C. where   D. what

Part 5 Applying

1. Enjoy the Idioms and Try to Use Them.

Every dog has its day.

这里的 day,是指 opportunities,即成功的机会;而 dog 则泛指那些地位低下、似乎没有出息的人。"士别三日,当刮目相看",落魄的人若发奋图强,总会有出头之日。

To laugh up one's sleeve

Sleeve 就是衣服袖子。To laugh up one's sleeve 从字面上来看是在袖子里笑,实际上也就是偷偷地笑。To laugh up one's sleeve 的真正意思就是偷偷地笑话某人,因为这个人有些可笑的地方,而他本人还没有发现。

To twist someone's arm

Twist 是拧,用力扭转的意思。一个人要是被别人把手臂拧到身后去那是很难受的。To twist someone's arm 这个习惯用语的意思也正是如此,它是指给某人施加压力,迫使他做你要他做的事。

2. Creative Writing — Features of Temples and Shrines in Japan

Directions: Please find out the representative architectural illustration in Kyoto, Japan. Write an essay about the architecture in English and introduce its time, specialties and style, etc. Print it in Paper A4 and send a copy to the teacher's e-mail box.

_____

_____

_____

_____

Lesson 6
The Classical Masterpiece in Ancient China

- **Part 1 Listening and Speaking**
  Topic   A Successful Trip

- **Part 2 Intensive Reading**
  Text A   Irrigation Works and Chain Bridges

- **Part 3 Extensive Reading**
  Text B   The Summer Palace

- **Part 4 Sentence Structures**
  Attributive Clause

- **Part 5 Applying**
  Enjoy These Proverbs and Try to Recite Them
  Creative Writing — on Zhaozhou Bridge

Lesson 6
The Classical Masterpiece in Ancient China

Part 1 Listening and Speaking

Topic A Successful Trip

Directions:　This part is to train your listening ability. The dialogue will be spoken twice. After the first reading, there will be a pause. During the pause, you must read after it and then listen to it again in order to put the sentences into Chinese.

A: John, I heard you went to Syria as part of an archaeological dig. Wow. That must have been quite an experience.

B: Yeah, and it was a colossal project. I think we had over 200 people working on it at the same time.

A: What could inspire someone to undertake such a large project?

B: Well, our investigation was an outgrowth of some buildings discovered last year. The architecture they previously uncovered had a very naturalistic style and some were quite austere. But the items we were searching for were ornaments, jewelry, and other exotic items.

A: Did you uncover much?

B: Yes. The civilization we were studying seemed to flourish. The project was a great victory, particularly for our director. He was so excited; we joked that he might organize a parade to celebrate.

A: That would have been quite a spectacle. I'm glad you had a successful trip. It's good to see you made it home safe and sound!

Chinese Versions

A: _____
B: _____
A: _____
B: _____
A: _____
B: _____
A: _____

Part 2 Intensive Reading

Text A

Irrigation Works and Chain Bridges

Along the long history of China, there are many brilliant architectural projects and miracles left and preserved which are worth admiring to modern people all over the world.

First of all, the Grand Canal was a giant irrigation project of ancient China. With a history of over 1,400 years and a length of 1,794 kilometers, it is one of the world's oldest canals and is the longest man-made river in the world, far surpassing the next two grand canals of the world: Suez and Panama Canal.

It goes from Tongxian County, Beijing in the north to Hangzhou in the south and connects five large rivers—the Haihe, Yellow, Huaihe, Qiantang and Yangtze River. In the late Spring and Autumn Period, it was first cut near Yangzhou, Jiangsu Province, to guide the waters of the Yangtze River to the north and expanded by Emperor Yangdi of Sui Dynasty during six years of furious construction to become what has been known as the Grand Canal. The Grand Canal was the major transport artery between north and south China during the Yuan, Ming and Qing Dynasties, contributing greatly to the economic and cultural exchange between north and south— a role denied to the large natural rivers that mostly flow from west to east.

Next to be mentioned is Dujiang Dam, which is a historical wonder of science and technology and the oldest water conservancy project still operating in the world. Dujiang Dam was designed and built to control flood over 2,200 years ago by Li Bing, a local governor of Shu State during the Warring States period. In addition to its irrigation and water control functions, the Dujiang Dam is also a beautiful tourist site listed as a "World Cultural Heritage" site by UNESCO. It consists of three main parts: the Fish Mouth Water—Dividing Dam, the Flying Sand Fence, and the Bottle-Neck Channel. The dam separates the turbulent river into outer and inner channels, the former serving as a drainage system to prevent floods and the latter leading water off to irrigate farmlands. Up to now, the works still play a significant role in agriculture.

Last come the bridges. China is the origin of many bridge forms: Marco Polo told of 12,000 bridges built of wood, stone, and iron near the ancient city of Kin-sai.

The most famous chain bridge is Luding Bridge in Ganzi, Sichuan province. Luding Bridge was built in 1705. It is an iron suspension bridge over Dadu River. It is 100 meters long, 28 meters wide and 14 meters high from the water surface. The whole bridge has 13 iron chains, each of them weighing 2.5 tons. Among them, nine parallel chains are tied to the two banks. The other four chains are suspended on the sides on the left and right, with two chains on either side to be the handrails.

All the works mentioned above embody the wisdom and ability of ancient Chinese people. Being proud of these cultural heritages, we should strive to carry our excellent national culture forward.

(504 words)

Images connected with Internet 欣赏相关视频截图

中国大运河沿岸的6省2市按行政区划分自南向北的省份分别为浙江、江苏、安徽、河南、山东和河北；城市为天津、北京。它的起点定在杭州拱宸桥，北端在通州。

http://www.nitchiewiki.com

http://www.go-passport.grolier.com

http://www.costaricapages.com

http://www.mnitcc.blogspot.com

http://www.doreandme.com

Word Study

Words	Pronunciation	Versions
irrigation	[ˌiri'geiʃən] n.	supplying dry land with water by means of ditches etc. 灌溉
irrigate	['irigeit] v.	to supply water to an area of land through pipes or channels so that crops will grow 灌溉
miracle	['mirəkl] n.	an event that appears inexplicable by the laws of nature 奇迹
preserve	[pri'zə:v] v.	to maintain in safety from injury, peril, or harm; protect 保护,保存
admire	[əd'maiə] v.	to regard with pleasure, wonder, and approval 钦佩,羡慕
grand	[grænd] adj.	large and impressive in size, scope, or extent; magnificent 主要的,极重要的
canal	[kə'næl] n.	an artificial waterway or artificially improved river used for travel, shipping, or irrigation 运河,沟渠
giant	['dʒaiənt] n.	a person or thing of great size 巨人,伟人
	adj.	marked by exceptionally great size, power, etc. 庞大的
surpass	[sə:'pɑ:s] v.	to be beyond the limit, powers, or capacity of 超越,胜过
expand	[iks'pænd] v.	to increase the size, volume, quantity, or scope of; enlarge 使膨胀,详述,扩张
furious	['fjuəriəs] adj.	with great energy, speed, or anger 激烈的,猛烈的,高速的
artery	['ɑ:təri] n.	a major route of transportation into which local routes flow 动脉,要道
contribute	[kən'tribju:t] v.	to help bring about a result; act as a factor 贡献,捐献
exchange	[iks'tʃeindʒ] v.	to give in return for something received; trade; interchange 交换,交易
conservancy	[kən'sə:vənsi] n.	conservation, especially of natural resources 水土保持,资源保护
turbulent	['tə:bjulənt] adj.	violently agitated or disturbed; tumultuous 狂暴的,吵闹的
drainage	['dreinidʒ] n.	the action or a method of draining; a system of drains 排水装置
suspension	[səs'penʃn] n.	[technical] a liquid with very small pieces of solid matter floating in it 悬浮液,悬浮 suspension bridge 索吊桥

handrail	['hændreil] n.	a narrow railing to be grasped with the hand for support 栏杆, 扶手
embody	[im'bɔdi] v.	to give a bodily form to; to represent in bodily or material form 具体化, 体现
strive	[straiv] v.	to exert much effort or energy; endeavor 努力, 奋斗

Proper Nouns

1. the Grand Canal 大运河
2. Suez Canal ['sju(:)iz] 苏伊士运河
3. Panama Canal [ˌpænə'mɑ:, 'pænəmə] 巴拿马运河
4. Emperor Yangdi of Sui Dynasty 隋炀帝
5. Dujiangyan Dam 都江堰
6. Luding Bridge: Luding Suspension Bridge 泸定桥(索吊桥)

Useful Expressions

❶ be worth doing 值得……
If something is worth doing, it is worth doing well. 如果事情值得做, 就值得做好。

❷ contribute (greatly) to... 为……做贡献
Lu Xun contributed greatly to Chinese literature. 鲁迅对中国文学做出了巨大的贡献。

❸ consist of 由……组成
One week consists of seven days. 一周有七天。

❹ separate ... into... 把……分成
A chemist can separate a medicine into its components. 化学家能把一种药物的各种成分分解出来。

❺ serve as... 作为, 用作, 供应……(饭菜)
This will serve as certification. 以此为凭。

❻ play a/an (important, significant) role in 扮演……角色
Mobile phone plays an important role in everyday life. 手机在日常生活中起着重要的作用。

❼ carry... forward 发扬光大
We should carry the revolutionary tradition forward. 我们应该把革命传统发扬光大。

Notes on Text A

❶ In the late Spring and Autumn Period, it was first cut near Yangzhou, Jiangsu Province, to guide the waters of the Yangtze River to the north and then expanded by Emperor Yangdi of Sui Dynasty during six years of furious construction to become what has been known as the Grand Canal.

> 分析　... was first cut 与 and then expanded 是由 and 连接的并列谓语；to guide the waters ... to the north 和 to become what has been known as the Grand Canal 分别为动词不定式短语，作目的状语，其中 what has been known as the Grand Canal 是由连接代词 what 引导的表语从句。

❷ The Grand Canal was the major transport artery between north and south China during the Yuan, Ming and Qing Dynasties, contributing greatly to the economic and cultural exchange between north and south — a role denied to the large natural rivers that mostly flow from west to east.

> 分析　contributing greatly to the economic and cultural exchange between north and south 是现在分词短语，做伴随状语；a role denied to the large natural rivers that mostly flow from west to east 是名词短语，做 transport artery 的同位语；denied to... 是过去分词短语，做名词 role 的后置定语；that mostly flow from west to east 是由关系代词 that 引导的限制性定语从句，修饰先行词 rivers。

❸ Next to be mentioned is the Dujiang Dam, which is a historical wonder of science and technology and the oldest water conservancy project still operating in the world.

> 分析　关系代词 which 引导非限制性定语从句，修饰先行词 Dujiang Dam；operating in the world 是现在分词短语，作 project 的后置定语。

❹ The dam separates the turbulent river into outer and inner channels, the former serving as a drainage system to prevent floods and the latter leading water off to irrigate farmlands.

> 分析　the former serving as... floods 和 the latter leading ... to irrigate farmlands 是两个现在分词短语的独立主格结构，在句中作伴随状语，其逻辑主语分别为 the former 和 the latter。

❺ the Spring and Autumn period 指中国古代春秋时期，从公元前770年到公元前475年间。

❻ Warring States period 指三国时期，从220年到280年。

❼ Kin-sai 是《马可波罗游记》一书中的地名，应为今天的浙江，杭州。

Part 3 Extensive Reading

Text B

The Summer Palace

The Summer Palace in Beijing—first built in 1750, largely destroyed in the war of 1860 and restored on its original foundations in 1886—is a masterpiece of Chinese landscape garden design. The natural landscape of hills and open water is combined with artificial features such as pavilions, halls, palaces, temples and bridges to form a harmonious ensemble of outstanding aesthetic value.

The Summer Palace, dominated mainly by Longevity Hill and Kunming Lake, covers an area of 2.9 square kilometers, three quarters of which are under water. Its 70,000 square meters of building space features a variety of palaces, gardens and other ancient-style architectural structures. Well known for its large and priceless collection of cultural relics, it was among the first group of historical and cultural heritage sites in China to be placed under special state protection.

The Summer Palace is a monument to classical Chinese architecture, in terms of both garden design and construction. Borrowing scenes from surrounding landscapes, it radiates not only the grandeur of an imperial garden but also the beauty of nature in a seamless combination that best illustrates the guiding principle of traditional Chinese garden design: "The works of men should match the works of Heaven".

In December 1998, UNESCO included the Summer Palace on its World Heritage List with the following comments: 1) The Summer Palace in Beijing is an outstanding expression of the creative art of Chinese landscape garden design, incorporating the works of humankind and nature in a harmonious whole; 2) The Summer Palace epitomizes the philosophy and practice of Chinese garden design, which played a key role in the development of this cultural form throughout the east; 3) The imperial Chinese garden, illustrated by the Summer Palace, is a potent symbol of one of the major world civilizations.

Word Study

Words	Pronunciation	Versions
destroy	[dis'trɔi] *v.*	to ruin completely; spoil; to tear down or break up; demolish 破坏, 毁坏, 消灭
restore	[ris'tɔː] *v.*	to bring back into existence or use; reestablish; to bring back to an original condition 修复, 重建
landscape	['lændskeip] *n.*	an expanse of scenery that can be seen in a single view 风景, 山水画, 地形, 前景

artificial	[ˌɑːtiˈfiʃəl] adj.	made by human beings; produced rather than natural 人造的,假的
ensemble	[əːnˈsɑːmbl] n.	[法] a unit or group of complementary parts that contribute to a single effect; a coordinated outfit or costume 全体,合唱曲
longevity	[lɔnˈdʒevəti] n.	great duration of life; length or duration of life 寿命
relic	[ˈrelik] n.	something that has survived the passage of time, especially an object or a custom whose original culture has disappeared 遗迹,废墟
seamless	[ˈsiːmləs] adj.	without a seam 无(接)缝的,无空隙的
radiate	[ˈreidieit] v.	to send out rays or wave; to manifest in a glowing manner 发光,辐射,流露
grandeur	[ˈgrændʒə] n.	the quality or condition of being grand; magnificence 庄严,伟大
epitomize	[iˈpitəmaiz] v.	to make an epitome of; sum up; to be a typical example of 摘要,概括,成为……缩影
potent	[ˈpəutənt] adj.	possessing inner or physical strength; powerful; having great control or authority 有力的,有效的

Proper Nouns

1. the Summer Palace 颐和园
2. Longevity Hill 万寿山
3. Kunming Lake 昆明湖
4. World Heritage List 世界文化遗产名录

Useful Expressions

❶ be combined with 与……相结合
In sum, theory must be combined with practice. 总之,理论必须同实践相结合。

❷ be well known for 因……而闻名
He is well known for fair dealing. 他以平等待人著称。

❸ in terms of 根据,按照,就……而言
Our firm is very strong in terms of manpower. 我们的公司在人力资源方面非常强大。

❹ incorporate ... into... 把……合并为……
We will incorporate your suggestion into the new plan. 我们将把你的建议编到新计划中去。

Notes on Text B

❶ The Summer Palace in Beijing—first built in 1750, largely destroyed in the war of 1860 and restored on its original foundations in 1886—is a masterpiece of Chinese landscape garden design.

分析 built in... destroyed in... and restored on 分别为过去分词短语，作定语，修饰the Summer Palace。

❷ The Summer Palace, dominated mainly by Longevity Hill and Kunming Lake, covers an area of 2.9 square kilometers, three quarters of which is under water.

分析 dominated mainly by ... lake 过去分词短语，做定语（也可以看作是插入语）；covers 谓语动词，意思是"覆盖"；three quarters of which is under water是由"名词+介词+which"引导的非限制性定语从句，修饰先行词area; three quarters of which 意思是"四分之三的面积"。

❸ Well known for its large and priceless collection of cultural relics, it was among the first group of historical and cultural heritage sites in China to be placed under special state protection.

分析 well known for... relics 是过去分词短语，作原因状语；to be placed... protection是动词不定式短语被动语态，作目的状语。

❹ Borrowing scenes from surrounding landscapes, it radiates not only the grandeur of an imperial garden but also the beauty of nature in a seamless combination that best illustrates the guiding principle of traditional Chinese garden design: "The works of men should match the works of Heaven".

分析 borrowing scenes from surrounding landscapes 现在分词短语，作原因状语；not only... but also... 并列连词，连接两个名词短语，作谓语动词radiates的宾语；that best illustrates... "The works of men should match the works of Heaven"是由关系代词that引导的限制性定语从句，修饰先行词combination。

Images connected with Internet 欣赏相关视频截图

1860年圆明园被烧毁之前，还没有颐和园这一称谓，当时的颐和园叫做清漪园(Garden of Clear Ripples)，也一同被烧毁。那时西方把圆明园称为Summer Palace。圆明园被烧毁28年后(1888年)，慈禧太后以筹措海军经费的名义动用3000万两白银重建清漪园，改称颐和园，其宏大华丽堪比圆明园，此后西方人始称颐和园为Summer Palace，而称被烧毁的圆明园为Old Summer Palace。将颐和园译为Summer Palace的说法一直延续至今。

Part 4 Sentence Structures

Attributive Clause 定语从句

在句子中充当定语的句子叫定语从句。它的作用相当于形容词。定语从句一般放在它所修饰的名词或代词之后。它所修饰的名词或代词叫做先行词。

定语从句由关系代词who, whom, whose, that, which, as 或关系副词where, when, why, 等引导;关系代词或关系副词既指代先行词,又在从句中充当句子成分,如:主语、宾语、表语、状语等。

1. Relative Pronouns 关系代词

(1) 关系代词who 在从句中作主语,其先行词是指人的名词或代词;who 可以代替whom,在从句中作宾语,口语中常常被省略,如:

He who loses money, loses much. 失金钱者,失去甚多。

He who loses a friend, loses much more. 失朋友者,失去更多。

He who loses faith, loses all. 失信言者,失去一切。

The boy who (whom) we met yesterday is called Tom. 昨天我们见到的那个男孩名叫汤姆。

(2) 关系代词whom 在从句中作宾语,其先行词是指人的名词或代词。口语中常省略,也可以用who 代替;whom 在从句中作介词宾语且介词前置时,whom 不能省略,也不能用who 代替,如:

Where are the people whom we are waiting for? 我们等的人在哪里?

The girl to whom you were just speaking is our guide. (The girl whom /who you were just speaking to is our guide.) 刚才你和她说话的那位姑娘是我们的导游。

(3) 关系代词whose 在从句中作定语,它可以是who 或which 的所有格,如:

The team is leaded by an Italian archaeologist, whose working experience is quite rich. 领队是位意大利考古学家,他的考古经验非常丰富。

The first teddy bear was born in 1902, whose name came from the American president. 第一只玩具熊诞生在1902年,它的名字来源于美国总统。

(4) 关系代词that 在从句中作主语或宾语,作宾语时常省略;其先行词是指人或指物的名词或代词。一般情况下 that 可以替代 who, whom 或 which, 如：

The house that (which) belongs to Mary is on sale today. 玛丽的房子今天出售。

Let me show you the novel that (which) I have just borrowed from the library. 你看这是我刚从图书馆借的小说。

I will interview the man that (whom) I phoned yesterday. 我要面试那位昨天打过电话的人。

I will interview the man that (who) phoned me yesterday. 我要面试那位昨天给我打电话的人。

(5) 关系代词 which 在从句中作主语或宾语，作宾语时常省略，作介词宾语时，介词前置，which 不能省略；其先行词是指物或事的名词，如：

The Great Wall has many gate ways which connect the main roads of North China. 长城有许多关口，这些关口与华北的主要道路相连。

This is the hotel which you will stay at. (可以说：This is the hotel at which you will stay. 或：This is the hotel (which /that) you will stay at. 或 This is the hotel where you will stay.) 这就是你入住的旅馆。

(6) 关系代词 as 在从句中作主语、宾语或表语，常以固定搭配 the same... as, such... as, as often happens, as has been said before, as mentioned above, as usual, as is well-known 等形式出现在句中，如：

This is the same bridge as we passed. 这和我们刚经过的那个桥一模一样。

Such buildings as mentioned above are out-of-date in this area nowadays.
上面提到的这种建筑物今天已经不多见了。

As everybody can see, great changes have taken place in China.
众所周知，中国已经发生了巨大的变化。

2. Relative Adverbs 关系副词

(1) 关系副词 where 在从句中作地点状语，其先行词是表示地点的名词，如：

This is the hotel where you will stay. 这就是你入住的旅馆。

The office where (at which) he works is not far from here. 他工作单位离这儿不远。

(2) 关系副词 when 在从句中作时间状语，其先行词是表示时间的名词。When 有时可省略，特别是当它的先行词是介词短语中的名词或在主句中作宾语时，如：

Do you ever have one of those days when everything goes wrong? 你经历过凡事都很糟糕的一天吗？

He came at a time when we needed help. 他总是在我们需要帮助的时候伸出援助之手。

He died on the day (when) his invention was successfully known to all. 他在他的发明公诸于世的那天去世了。

(3) 关系副词 why 在从句中作原因状语，其先行词有 excuse, explanation, reason 等，如：

Do you know the reason why she didn't come to the party. 你知道她为什么没参加晚会吗？

This is the explanation why I have changed our original plan. 这是我对改变原计划的解释。

(4) 关系副词 how 在从句中作方式状语，其先行词是 way，如：

This is the way how you treat it. 你就这样处理它。

(5) 关系副词 that 在从句中可以作时间、地点、原因、方式等状语，其先行词多为表示时间、地点、原因、方式等的名词，that 常常被省略，如：

Spring is the time (that, when, in which) people go sightseeing. 春天是人们观光的季节。

Do you know any place (that) we should enjoy ourselves safe and sound? 你知道我们能到哪里去痛痛快快的玩吗？

3. Restrictive and Non-restrictive Attributive Clauses
限制性定语从句和非限制性定语从句

定语从句分限制性定语从句和非限制性定语从句两种。

(1) 限制性定语从句是先行词不可缺少的定语。如果将这种从句省去，主句的意思就会不完整或不明确。它和主句的关系十分密切，不可用逗号与主句隔开。其关系词有时可省略，如：

This is the most beautiful architecture that (which) we have ever seen. 这是我们见过的最漂亮的建筑。

The coat which/that she washed this morning is still wet. 她早晨洗的外套还没晾干。

He is one of the designers who comes from America. 他是一位来自美国的设计师。

(2) 非限制性定语从句是先行词（或者短语、句子等）的附加说明部分，如果省略，主句的意思仍然很完整，它和主句之间通常用逗号隔开。非限制性定语从句不能用 that 或 why 引导，关系词不能省略。此外，非限制性定语从句在语义上有时相当于一个并列句，有时相当于一个状语从句，如：

Next to be mentioned is the Dujiang Dam, which is a historical wonder of science and technology and the oldest water conservancy project still operating in the world. 下面提到的是都江堰，它是一个史上科技奇迹，也是世界上仍在使用的最为古老的水利工程。

He is recovering quickly, which I am glad to hear. 他康复得很快，我听了很高兴。

She wants to write an article, which will attract public attention to this matter. 她想写一篇文章，以便引起公众对这件事的注意。

Abraham Lincoln, who led the United States through these years, was shot on April 14, 1865. 亚伯拉罕·林肯领导美国人民度过这些年，却在1865年4月14日被人枪杀。

He didn't come, as I had expected. 正如我所预料的那样，他没有来。

His father, who is now living in Beijing, is a garden designer. 他父亲住在北京，是个园林设计师。

Practice

Choose the best answer to complete each sentence.

1) The place _____ interested me most was the Children's Palace.
   A. which   B. where   C. what   D. in which

2) Do you know the man _____?
   A. whom I spoke   B. to who speak   C. I spoke to   D. that I spoke

3) This is the hotel _____ last month.
   A. which they stayed   B. at that they stayed
   C. where they stayed at   D. where they stayed

4) Do you know the year _____ the Chinese Communist Party was founded?
   A. which   B. that   C. when   D. on which

5) That is the day _____ I'll never forget.
   A. which   B. on which   C. in which   D. when

6) The factory _____ we'll visit next week is not far from here.
   A. where   B. to which   C. which   D. in which

7) Great changes have taken place since then in the factory _____ we are working.
   A. where   B. that   C. which   D. there

8) This is one of the best films _____.
   A. that have been shown this year   B. that have shown
   C. that has been shown this year   D. that you talked

9) Can you lend me the book _____ the other day?
   A. about which you talked   B. which you talked
   C. about that you talked   D. that you talked

10) The pen _____ he is writing is mine.
    A. with which   B. in which   C. on which   D. by which

Part 5 Applying

1. Enjoy These Proverbs and Try to Recite Them.

1. Man proposes, God disposes.
   谋事在人，成事在天。
2. Ambition is to life just what steam is to the locomotive.
   壮志之于人生，恰似蒸汽之于火车头。
3. No pains, no gains.
   一分耕耘，一分收获。
4. While there is life, there is hope.
   留得青山在，不怕没柴烧。
5. If a thing is worth doing it is worth doing well.
   如果事情值得做，就值得做好。

2. Creative Writing — on Zhaozhou Bridge

赵州桥位于河北赵县洨河上，它是世界上现存最早、保存最好的巨大石拱桥。被誉为"华北四宝之一"。桥长64.40米，跨径37.02米，高7.23米，是当今世界上跨径最大、建造最早的单孔敞肩型石拱桥。因桥两端肩部各有二个小孔，不是实的，故称敞肩型，这是世界造桥史上的一个奇迹（没有小拱的称为满肩或实肩型）。桥上有很多的东西，类型众多，丰富多彩。著名桥梁专家茅以升说，先不管桥的内部结构，仅就它能够存在1300多年就说明了一切。

图片来源：http://img.etu6.com/remote/20100413/etu6-201004131132191500.jpg

Directions: Write an article to introduce the most famous old bridge in China—Zhaozhou Bridge. You have got some information needed above, and try to make a speech in your own words. Print the essay out in Paper A4 and send it to your English teacher's e-mail box.

Lesson 7
The Ancient Architecture in the World

- **Part 1 Listening and Speaking**

  Topic   Gratitude on a Trip

- **Part 2 Intensive Reading**

  Text A   The Egyptian Pyramid

- **Part 3 Extensive Reading**

  Text B   Famous Ancient Relics in the World

- **Part 4 Sentence Structures**

  Adverbial Clause

- **Part 5 Applying**

  Enjoy the Famous Sayings and Try to Read Them Aloud

  Creative Writing — on an Ancient Relic in Egypt

Lesson 7
The Ancient Architecture in the World

Part 1 Listening and Speaking

Topic Gratitude on a Trip

Directions:　This part is to train your listening ability. The dialogue will be spoken twice. After the first reading, there will be a pause. During the pause, you must read after it and then listen to it again in order to put the sentences into Chinese.

A: Hi, Mr. Li. I'm here to express my appreciation for your help all these days. I really enjoyed my stay here.

B: You're welcome.

A: I'm returning to America today. So I thank you for all the time you've spend on my account during my stay here.

B: Don't mention it. I was pleased to be of assistance.

A: If there's anything I can help you with in the future, please let me know.

B: Thank you. Have a safe trip home.

A: Sure. Take care.

Chinese Versions

A: _____
B: _____
A: _____
B: _____
A: _____
B: _____
A: _____

Part 2 Intensive Reading

Text A

The Egyptian Pyramid

The pyramids of Egypt fascinated travelers and conquerors in ancient times and continue to inspire wonder in the tourists, mathematicians, and archeologists who visit, explore, measure and describe them.

The largest and most famous of all the pyramids, the Great Pyramid at Giza, was built by Snafu's son, Khufu, known also as Cheops, the latter Greek form of his name. The pyramid's base covered over 13 acres and its sides rose at an angle of 51 degrees 52 minutes and were over 755 feet long. It originally stood over 481 feet high; today it is 450 feet high. Scientists estimate that its stone blocks average over two tons apiece, with the largest weighing as much as fifteen tons each. Two other major pyramids were built at Giza, for Khufu's son, King Khafre, and a successor of Khafre, Menkaure. Menkaure, also located at Giza, is the famous Sphinx, a massive statue of a lion with a human head, carved during the time of Khafre.

There has been speculation about pyramid construction. Egyptians had copper tools such as chisels, drills, and saws that may have been used to cut the relatively soft stone. The hard granite, used for burial chamber walls and some of the exterior casing, would have posed a more difficult problem. Workmen may have used an abrasive powder, such as sand, with the drills and saws. Knowledge of astronomy was necessary to orient the pyramids to the cardinal points, and water-filled trenches probably were used to level the perimeter. A tomb painting of a colossal statue being moved shows how huge stone blocks were moved on sledges over ground first made slippery by liquid. The blocks were then brought up ramps to their positions in the pyramid. Finally, the outer layer of casing stones was finished from the top down and the ramps dismantled as the work was completed.

Pyramid building was at its height from the Fourth through the Sixth Dynasties. Smaller pyramids continued to be built for more than one thousand years. Scores of them have been discovered, but the remains of others are probably still buried under the sand. As it became clear that the pyramids did not provide protection for the mummified bodies of the kings but were obvious targets for grave robbers, later kings were buried in hidden tombs cut into rock cliffs. Although the magnificent pyramids did not protect the bodies of the Egyptian kings who built them, the pyramids have served to keep the names and stories of those kings alive to this day.

(422words)

Images connected with Internet 欣赏相关视频截图

Sphinx

Khufu Pyramid

Step Pyramid

Red Pyramid

Pyramid of the Sun

Pyramid of the Moon

Word Study

Words	Pronunciation	Versions
pyramid	['pirəmid] n.	a polyhedron having a polygonal base and triangular sides with a common vertex 金字塔
conqueror	['kɔŋkərə] n.	someone who is victorious by force of arms 征服者,胜利者
archeologist	[ˌɑːki'ɔlədʒist] n.	an anthropologist who studies prehistoric 考古学家
acre	['eikə] n.	a unit of area (4840 square yards) 英亩
apiece	[ə'piːs] adv.	to or from every one of two or more (considered individually) 每个,各
sphinx	[sfiŋks] n.	one large stone statue with the body of a lion and the head of a man that was built by the ancient Egyptians 狮身人面像
massive	['mæsiv] adj.	imposing in size or bulk or solidity 巨大的
speculation	[ˌspekju'leiʃən] n.	a message expressing an opinion based on incomplete evidence 沉思,推测,投机
chisel	['tʃizl] n.	a tool with a sharp flat edge at the end, used for shaping wood, stone or metal 凿子

drill	[drɪl]	n.	a tool or machine with a pointed end for making holes 钻,钻头,钻床,钻机
burial	['berɪəl]	n.	the ritual placing of a corpse in a grave 埋葬
bury	['berɪ]	v.	to put someone who has died in a grave
chamber	['tʃeɪmbə]	n.	a room used for a special purpose, especially an unpleasant one 密室,房间
abrasive	[ə'breɪsɪv]	n.	a substance that abrades or wears down 研磨剂
		adj.	causing abrasion 磨平的
astronomy	[ə'strɒnəmɪ]	n.	the branch of physics that studies celestial bodies and the universe as a whole 天文学
orient	['ɔːrɪənt]	v.	determine one's position with reference to another point 使朝向,以……为方向,使(尤指教堂)朝东建造
cardinal	['kɑːdɪnəl]	adj.	serving as an essential component 主要的,最重要的
trench	[trentʃ]	n.	a ditch dug as a fortification having a parapet of the excavated earth 沟,沟渠
perimeter	[pə'rɪmɪtə]	n.	the boundary line or the area immediately inside the boundar 周围,边缘
ramp	[ræmp]	n.	a slope that has been built to connect two places that are at different levels 坡道 off-/on-ramp 下/上坡
dismantle	[dɪs'mæntl]	v.	tear down so as to make flat with the ground; take apart into its constituent pieces 拆除……的设备,分解
mummify	['mʌmɪfaɪ]	v.	remove the organs and dry out (a dead body) in order to preserve it 将(尸体)制成木乃伊
mummified	['mʌmɪfaɪd]	adj.	木乃伊化的

Proper Nouns

1. Egypt 埃及

   a republic in northeastern Africa known as the United Arab Republic until 1971; site of an ancient civilization that flourished from 2600 to 30 BC

2. Egyptian 埃及的,埃及人

of or relating to or characteristic of Egypt or its people or their language

3. Giza 吉萨省(埃及省份,位于开罗西南面)

   an ancient Egyptian city on the west bank of the Nile opposite Cairo; site of three Great Pyramids and the Sphinx

4. Khufu 胡夫(古埃及第四王朝法老,在位期间下令修建了最大的金字塔)

   Egyptian Pharaoh of the 27th century BC who commissioned the Great Pyramid at Giza

5. Cheops 基奥普斯 (Khufu 的希腊名)

Useful Expressions

❶ locate at 位于,座落在……(附近)

Both companies locate at the neighborhood of our factory. 这两家公司都位于我们工厂邻边。

❷ be used to 被用来

Lasers can be used to perform operations nowadays. 现在激光可以用来做手术。

❸ be necessary to 为……所必需

To gain that which is worth having, it may be necessary to lose anything else. 要获得值得拥有的东西,可能必须失去所有其他的。

❹ bring up (to) 提高,达到

The road will bring you up to the top of the cliff. 这条道路将使你们到达悬崖的顶部。

❺ scores of 大量,许多

Scores of guests had been trapped in their rooms—too terrified to move. 许多客人被困在房间里,害怕得不能动。

❻ provide for 供给 (为……作准备,规定,考虑到)

We must provide for the future. 我们必须为将来做好准备。

❼ serve to 用来 (足以……)……

This success will only serve to spur her on. 这个成绩只会鼓舞她继续前进。

❽ to this day 至今,迄今,直到现在

Her secret remains untold to this day. 她的秘密至今仍未透露。

Notes on Text A

❶ Snefru 斯奈夫鲁 第四王朝的首代国王,他建造了美都姆(Maydum) 的多级金字塔,然后又加以改筑,使其形成第一座真正的金字塔。

❷ King Khafre 卡夫拉国王

❸ Menkaure 孟考拉 第三座金字塔是安葬胡夫孙子孟考拉(Menkaure)的金字塔，建于公元前2600年左右。由于王朝的衰落，金字塔的建筑也开始衰落。

❹ The largest and most famous of all the pyramids, the Great Pyramid at Giza, was built by Snefru's son, Khufu, known also as Cheops, the latter Greek form of his name.

分析 The largest and most famous of all the pyramids, ... 是形容词短语，作 the Great Pyramid 的同位语，置于句首表示加强语气；known also as Cheops 是过去分词短语，在句中作伴随状语。the latter Greek form of his name 是省略句型，主语 the latter 后省略了谓语动词 was。

❺ The pyramid's base covered over 13 acres and its sides rose at an angle of 51 degrees 52 minutes and were over 755 feet long. It originally stood over 481 feet high; today it is 450 feet high.

分析 该句为并列句，由第一个 and 连接；第二个 and 连接两个并列谓语 rose 和 were。

1 acre means 4047 square meters；1 meter means 3.280840 feet.

❻ Knowledge of astronomy was necessary to orient the pyramids to the cardinal points, and water-filled trenches probably were used to level the perimeter.

分析 该句为并列句，由 and 连接；to orient the pyramids to the cardinal points 是动词不定式短语作目的状语；were used to level the perimeter. 意为：被用来（使金字塔的）底座呈水平状。

Part 3 Extensive Reading

Text B

Famous Ancient Relics in the World

Angkor Wat

Built between the ninth and the thirteenth centuries by a succession of twelve Khmer kings, Angkor spreads over 120 square miles in Southeast Asia and includes scores of major architectural sites. In 802, when construction began on Angkor Wat, with wealth from rice and trade, Jayavarman II took the throne, initiating an unparalleled period of artistic and architectural achievement, exemplified in the fabled ruins of Angkor, center of the ancient empire. Among the amazing pyramid and mandala shaped shrines preserved in the jungles of Cambodia, is Angkor Wat, the world's largest temple, an extraordinarily complex structure filled with iconographic detail and religious symbolism. Today, many countries continue efforts to conserve and restore the temples, which have been inaccessible until recently.

Maya City

Temple-pyramids were the most striking feature of a Classic Maya city. They were built from hand-cut limestone blocks and towered over all surrounding structures. Although the temples themselves usually contained one or more rooms, the rooms were so narrow that they could only have been used on ceremonial occasions not meant for public consumption. The alignments of ceremonial structures could be significant.

Maya City (Figure 1)

Typical Maya architectural features included the corbel vault and the roof comb. The corbel vault has no keystone, as European arches do, making the Maya vault appear more like a narrow triangle than an archway. It has been suggested that this unusual form exists because the Maya never mastered keystone technology. Others suggest that the lack of keystone was deliberate: the Maya vault always had nine stone layers, representing the nine layers of the Underworld. A keystone would have created a tenth layer, outside the Maya cosmology.

Perhaps Maya architects didn't feel the temples were grand enough, and so added an upper extension. The roof comb was always highly decorated with painted stucco reliefs, as was the temple facade. Equally decorated were the doorways, doorjambs and facades of many other Maya structures, which were ornamented with heavy carving in stone or wood.

Taj Mahal (Figure 2)

Almost everybody has read about the Taj Mahal in India. It is one of the most beautiful buildings in the world. Over three hundred years ago Shah Jehan built the Taj Mahal as a tomb for his wife.

Shah Jehan wanted his wife's tomb to be perfect. He did not care about time or money. He brought together workmen from all Asia. Altogether, over 20,000 men worked on the building and it took them over seventeen years to finish it.

The building rests on a platform of red sandstone. Four thin white towers rise from the corners of the platform. A large dome rests at the centre of the building. Around this large dome there are four smaller ones. The building is made of fine white and colored marbles. It has eight sides and many arches.

A beautiful garden surrounds the Taj Mahal. The green trees make the marble look even whiter. In front of the main entrance to the building there is a long, narrow pool.

(501 words)

Images connected with Internet 欣赏相关视频截图

full view of Angkor Wat

Angkor Wat

the Goddess Relief on Angkor Wat's Gallery

Word Study

Words	Pronunciation	Versions
succession	[sək'seʃən] n.	a following of one thing after another in time; acquisition of property by descent or by will 连续, 继承权, 继位
throne	[θrəun] n.	the chair of state of a monarch, bishop, etc. 王座, 君主
initiate	[i'niʃieit] v./n./adj.	v. to arrange for something important to start, such as an official process or a new plan 开始, 传授 n. start 入会, 开始 adj. newly added 新加入的
exemplify	[ig'zemplifai] v.	to be a very typical example of something 例证, 例示, 以……为典型
fabled	['feibld] adj.	celebrated in fable or legend 虚构的
jungle	['dʒʌŋgl] n.	an impenetrable equatorial forest 丛林, 密林
shrine	[ʃrain] n./v.	n. a place of worship hallowed by association with some sacred thing or person 圣地, 神龛 v. to place sth to a place of worship 将……置于神龛内
iconographic	[ai,kɔnə'græfik] adj.	of the way that a particular people, religious or political group etc represent ideas in pictures or images 象征性的
symbolism	['simbəlizəm] n.	a system of symbols and symbolic representations (尤指文艺中的)象征主义, 象征手法
conserve	[kən'sə:v] v.	keep in safety and protect from harm, decay, loss, or destruction 保存, 保全
inaccessible	[,inæk'sesəbl] adj.	capable of being reached only with great difficulty or not at all 难接近的, 难达到的, 难以达成的
ceremonial	[,seri'məunjəl] adj.	marked by pomp or ceremony or formality 正式的
consumption	[kən'sʌmpʃən] n.	(economics) the utilization of economic goods to satisfy needs or in manufacturing 消费

alignment	[əˈlainmənt] n.	the act of adjusting or aligning the parts of a device in relation to each other 调整(成直线,准线) 定向,直线性
comb	[kəum] n.	a flat device with narrow pointed teeth on one edge 梳子
keystone	[ˈkiːstəun] n.	a central cohesive source of support and stability; the central building block at the top of an arch or vault 拱心石,楔石
corbel	[ˈkɔːbəl] n.	(architecture) a triangular bracket of brick or stone (usually of slight extent) 承材,枕梁
archway	[ˈɑːtʃwei] n.	a passageway under a curved masonry construction 拱门,拱道
deliberate	[diˈlibəreit] adj.	intended or planned 故意的
	v.	to think about something very carefully 深思熟虑
cosmology	[kɔzˈmɔlədʒi] n.	the metaphysical study of the origin and nature of the universe 宇宙学
stucco	[ˈstʌkəu] n.	a plaster now made mostly from Portland cement and sand and lime; applied while soft to cover exterior walls or surfaces 灰泥
relief	[riˈliːf] n.	sculpture consisting of shapes carved on a surface so as to stand out from the surrounding background 浮雕
facade	[fəˈsɑːd] n.	the front of a building, especially a large and important one（建筑物的）正面,门面,外观
doorjamb	[ˈdɔːdʒæm] n.	a jamb for a door especially American English, a doorpost 大门柱
marble	[ˈmɑːbl] n.	a type of hard rock that becomes smooth when it is polished, and is used for making buildings, statues, etc. 大理石

Proper Nouns

1. Khmer [kəˈmɛə] the Mon-Khmer language spoken in Cambodia; a native or inhabitant of Cambodia 谷美尔人,谷美尔语
2. Cambodia [kæmˈbəudiə] a nation in southeastern Asia; was part of Indochina under French rule until 1946 柬埔寨 (亚洲)
3. Maya [ˈmɑːjə] a member of an American Indian people of Yucatan and Belize and Guatemala who had a culture (which reached its peak between AD 300 and

900) characterized by outstanding architecture and pottery and astronomy 玛雅人,玛雅语,玛雅人的,玛雅语的

4. mandala ['mændələ] any of various geometric designs (usually circular) symbolizing the universe; used chiefly in Hinduism and Buddhism as an aid to meditation 【宗】曼荼罗,坛场(佛教和印度教修法地方的圆形或方形标记。一些东方国家把佛、菩萨像画在纸帛上,亦称曼荼罗。)

5. Taj Mahal ['tɑːdʒməˈhɑːl] beautiful mausoleum at Agra built by the Mogul emperor Shah Jahan (completed in 1649) in memory of his favorite wife 泰吉马哈尔陵(亦译泰姬陵,印度阿格拉的一座大理石陵墓,由17世纪莫卧儿帝国皇帝 Shah Jahan 为爱妃所建。)

Useful Expressions

❶ a succession of 一连串的

There is a succession of rainy days here. 这里一连好几天都在下雨。

❷ take the throne 登基为王,加冕为王

Because Japan does not allow women to take the throne, Sayako must leave the imperial family when she marries Kuroda, a childhood friend of her brother. 日本不允许女性继承王位,因此,纪宫公主嫁给黑田庆树后必须离开皇室,黑田庆树是她哥哥儿时的朋友。

❸ more... than 与其说……倒不如说……,不是……而是……

Hearing the loud noise, the boy was more surprised than frightened. 这个男孩听到一声巨响后,没有害怕而是很吃惊。

❹ decorate with 以……来装饰

People usually decorate the house with paper cuts. 人们通常用剪纸装饰房子。

❺ care about 关心

I don't care about the price, so long as the car is in good condition. 我不计较价钱,只要车况好用就行了。

Notes on Text B

❶ Angkor ['æŋkə:] 吴哥

柬埔寨西北部的游览、考古胜地和古都;吴哥古迹包括 Angkor Thom 吴哥通王城和 Angkor Wat 吴哥窟。

❷ Jayavarman 贾亚瓦曼

位于吴哥城中央的拜云寺,建于12世纪,是当时国王贾亚瓦曼(Jayavarman)七世皇家

寺庙。其中矗立着37座石塔,每座石塔上皆刻有4张微笑脸,据说这些笑脸是贾亚瓦曼七世和佛祖脸混合体为蓝本雕刻。Jayavarman II 苏耶跋摩二世,柬埔寨吴哥王朝创立者。

❸ Shah Jehan 沙贾汗

据记载,印度皇帝Shah Jehan(沙贾汗)爱上了美丽的慕塔芝玛。于是他献上一颗价值一万多卢比的钻石给慕塔芝玛的父亲,首相Jehangir(贾汗季),请求将慕塔芝玛嫁给他。

❹ In 802, when construction began on Angkor Wat, with wealth from rice and trade, Jayavarman II took the throne, initiating an unparalleled period of artistic and architectural achievement, exemplified in the fabled ruins of Angkor, center of the ancient empire.

分析 1) with... 是介词短语做伴随状语；2) initiating... 是现在分词短语以及 exemplified... 是过去分词短语,还有center of ... 是名词短语作Angkor Wat的非限制性定语。

❺ Typical Maya architectural features included the corbel vault and the roof comb.

分析 vault 意为:基于拱架结构主体上的屋顶。the corbel vault 叠涩拱顶,叠涩拱顶的构造是从两堵或四堵墙上建起一系列起悬臂梁(cantilever)作用的挑托(corbel),直至挑托相互连接,形成拱顶。the corbel arch 叠涩拱,指用砖石层层堆叠向内收最终在中线合拢成的拱。叠涩拱技术起源甚早,在玛雅、古希腊等古代文明均有所发现。位于爱尔兰东北部,大约建于公元前3000年的纽格莱奇史前坟墓,亦发现一个完整无缺的叠涩拱,用以支撑主要墓室的屋顶。roof comb 意为:条脊。指玛雅金字塔那不朽的穹顶建筑结构,可以译作"鱼骨结构"。"The roof comb is the structure that tops a pyramid in monumental Mesoamerican architecture."

图示:左:真拱 右:叠涩拱

Part 4 Sentence Structures

Adverbial Clause 状语从句

修饰主句的动词、形容词或副词的句子叫状语从句。通常由从属连词引导,按其意义和作用可分为时间、地点、条件、原因、让步、目的、结果、方式和比较状语从句等。

1. Adverbial Clause of Time 时间状语从句

引导状语从句的连词有 after, as, before, once, since, till, (not) until, when, whenever (no matter when), while, as long as, as soon as, no sooner... than, hardly (scarcely, barely)... when, immediately, the moment, the minute, every time 等。

when 可表示一段时间,也可表示时间的某个点;while 只能表示一段时间,且常用进行时态;as 表示"一边……一边……",如:

no sooner... than, hardly(scarcely, barely)... when,表示"刚要……就……",一般主句用过去完成时,从句用一般过去时,如 no sooner 或 hardly 在句首,要用倒装语序,如:

Finally, the outer layer of casing stones was finished from the top down and the ramps dismantled as the work was completed. 最后,墓穴的外墙是自上而下完成的,完工后再把坡道拆除。

In 802, when construction began on Angkor Wat, with wealth from rice and trade, Jayavarman Ⅱ took the throne. 公元802年开始修建吴哥窟的时候,阇耶跋摩(又称贾亚瓦曼或苏耶跋摩)二世凭借做大米和货物贸易发了财并夺得王位。

As you look at yourself in a mirror, you'll see an identical image of yourself. 当你在镜子中看自己的时候,你会看到自己的相同的图像。

When I got to the airport, I suddenly remembered that I had left the ticket behind. 当我到达机场的时候,突然想起我未带机票。

I was about to leave, when something occurred which attracted my attention. 我刚要走,突然发生了一些事情吸引了我的注意力。

Whenever we have difficulty, he'll come to help us. 无论我们什么时候有困难,他都会帮助我们。

No sooner had I opened the door than the telephone rang. 我刚一开门电话铃就响了。

I'll tell you about it the moment you come. 你一来我就告诉你。

I got in touch with him immediately I received his letter. 我刚收到他的信,我们就取得了联系。

2. Adverbial Clause of Place 地点状语从句

引导地点状语从句的连词有 where 和 wherever,如:

I will stand where I can see the singers clearly. 我要站在能清楚看见这些歌手的地方。

Wherever you go, whatever you do, I'll be right here waiting for you. 无论你在哪儿,无论你做什么,我都会在这儿等你。

3. Adverbial Clause of Condition 条件状语从句(真实条件从句)

引导条件状语从句的连词有 if(如果), unless(除非), so long as(只要), provided that(假如,假设), in case that(万一)等,如:

I will not go to her party if she doesn't invite me. 如果她不邀请我,我就不去参加宴会。

I will not go to her party unless she invites me. 除非她邀请我,否则我不去参加宴会。

You may borrow this book as long as you promise to give it back. 只要你答应归还,你就可以把这书借走。

You may keep the book a further week provided (that) no one else requires it. 倘若这本书没有其他人想借的话,你可以再续借一个礼拜。

4. Adverbial Clause of Cause 原因状语从句

引导原因状语从句的连词有 because(因为), as(由于), since(既然), now that(既然), in that(以为)等,如:

Now (that) you are grown up, you should not rely on your parents. 既然你长大了,就不应该依靠你的父母。

Because he was ill he didn't attend the meeting last week. 上周他病了,没来开会。

5. Adverbial Clause of Concession 让步状语从句

引导让步状语从句的连词有 even if（即使）, though/although（虽然……但是）, even though（即使）, while（尽管）, no matter what/how/which（不管什么/怎样/哪个）, however（无论多么）, whatever（无论什么）, whichever（无论哪个）, whether... or（无论……还是）, as（虽然）等,如:

Although the magnificent pyramids did not protect the bodies of the Egyptian kings who built them, the pyramids have served to keep the names and stories of those kings alive to this day. 尽管宏伟的金字塔没有把建造金字塔的埃及国王的遗体保存下来,但是它确把这些国王的名字以及他们的故事流传至今。

Rich as he is, he is not happy. 虽然他很富,但他并不幸福。

Try as he would, he could not lift the rock. 他虽然尽了最大努力,仍不能搬动那块石头。

He will not give up smoking even though the doctor advises him to. 即使医生建议他戒烟,他也不听。

Whatever the consequence may be, I will be on your side. 无论结果怎样,我都会站在你的一边。

However hard she tried to explain, nobody trusted her. 无论她怎么解释,都没有人相信她。

It has been the same result, whichever way you do it. 无论你用哪种方式去做,结果都是一样的。

6. Adverbial Clause of Result 结果状语从句

引导结果状语从句的连词有 so that, so... that, such... that 等,如:

The book is so interesting that I want to read it once more. 这本书太有趣了,以致于我想再读一遍。

She is such an open-minded person that we all like her. 她是那样一个热情开朗的人，我们都很喜欢她。

7. Adverbial clause of Purpose 目的状语从句

引导目的状语从句的连词有 so that (为了，以便), in order that (为了，以便), for fear that (唯恐，害怕), lest (以免), in case (以防，以免), 如：

They set out early so that they can arrive in time. 他们动身很早以便能及时到达。

Please remember to take your umbrella in case it rains. 记得带伞，以防下雨。

8. Adverbial Clause of Manner 方式状语从句

引导方式状语从句的连词有 as (按照……，像……那样), just as (正如……一样), as if (好像，似乎), 如：

You talk as if you had really been there. 你谈话的样子好像你去过那儿。

It looks as if it is going to rain. 看起来好像要下雨了。

9. Adverbial Clause of Comparation 比较状语从句

引导方式状语从句的连词有 as... as, not so/as... as , 比较级+than, so much/a lot more than, no more... than, not more... than, less... than, the more... the more；从句部分常有省略现象，如：

The film was not so exciting as we expected. 这部电影没有想象的那样令人兴奋。

The history of nursing is as old as the history of man. 护理的历史和人类的历史一样古老。

She looks much younger than she is. 她看上去年轻得多。

Jack is not more frightened than Mike is. 杰克不像迈克那样害怕。

The more I see him, the less I like him. 我看到他的次数越多越不喜欢他。

Practice

Choose the best answer to each sentence.

1) You can't get a driver's license _____ you are at least sixteen years old.
   A. if   B. unless   C. when   D. though

2) The young man lost his job last month, but it wasn't long _____ he found a new position in my company.
   A. before   B. while   C. as   D. after

3) _____ you have any questions or needs, please contact the manager after 5:00 p.m. on weekdays.
   A. Because   B. Where   C. If   D. Though

4) _____ Susan gets onto the top of a tall building, she will feel very much frightened.
   A. Now that B. Even though C. Every time D. Only if

5) When he went out, he would wear sunglasses _____ nobody would recognize him.
   A. so that B. now that C. as though D. in case

6) The hotel _____ during the vacation was rather poorly managed.
   A. as I stayed B. where I stayed C. which I stayed D. what I stayed

7) Li Lei didn't meet the famous American professor _____ he was on holiday in America last year.
   A. unless B. until C. if D. whether

8) She didn't go to the cinema last night, _____ she had to finish her term paper.
   A. as B. if C. till D. though

9) No sooner had they got off the train _____ it started moving.
   A. when B. than C. then D. after

10) It is very important for the strong man to know that _____ strong he is, he cannot be the strongest.
    A. whatever B. whenever C. whichever D. however

Part 5 Applying

1. **Enjoy the Famous Sayings and Try to Read Them Aloud.**

Miracles sometimes occur, but one has to work terribly for them. ——D Weizmann
奇迹有时候是会发生的,但是你得为之拼命的努力。 ——魏茨曼

Between the ideal and the reality, between the motion and the act, fall the shadows.
—— Eliot
理想与现实之间,动机与行为之间,总有一道阴影。 ——爱略特

Success often depends upon knowing how long it will take to succeed. ——Montesquieu
成功常常取决于知道需要多久才能成功。 ——孟德斯鸠

2. Creative Writing — on an Ancient Relic in Egypt

Directions: Make a speech on an ancient relic in Egypt in English and show some pictures related to this ancient relic through the Internet. Print the speech out in Paper A4 and send the speech to your English teacher via an E-mail.

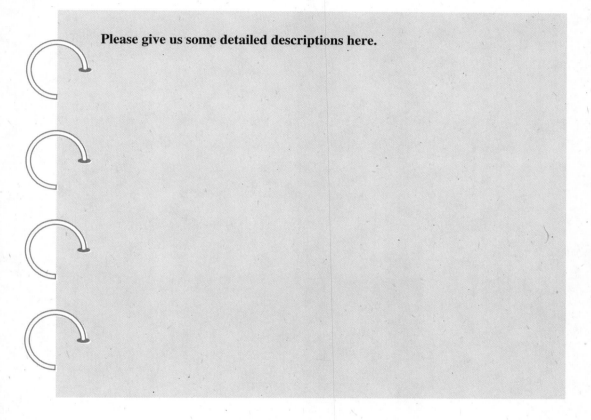

Please give us some detailed descriptions here.

Lesson 8
Ecology and Architecture

- **Part 1 Listening and Speaking**
  Topic   Asking a Volunteer for Help

- **Part 2 Intensive Reading**
  Text A   Better City, Better Life

- **Part 3 Extensive Reading**
  Text B   Konarka Power Plastic

- **Part 4 Sentence Structures**
  Subjunctive Mood

- **Part 5 Applying**
  Enjoy the Famous Sayings and Try to Recite Them
  Creative Writing — on Building Materials

Lesson 8
Ecology and Architecture

Part 1 Listening and Speaking

Topic Asking a Volunteer for Help

Directions: This part is to train your listening ability. The dialogue will be spoken twice. After the first reading, there will be a pause. During the pause, you must read after it and then listen to it again in order to put the sentences into Chinese.

A: Good afternoon, madam. Is there anything that I can do for you?
B: Oh, yes, please. Could you help me carry my luggage to the gate over there?
A: No problem.
B: It's really nice of you.
A: My pleasure.
B: By the way, could you tell me how I can get to the airport?
A: You can take the bus to get there. The bus station is just around the corner.
B: How long will it take me to get there?
A: About an hour, I think.
B: My flight is at four o'clock. I'm afraid I won't be able to make it.
A: If that's the case, I'd suggest you take a taxi.
B: Sounds right.
A: If you don't mind, you may stay here with your luggage while I go and get you a taxi.
B: That would be very nice of you. Thank you very much.
A: You are welcome.

Chinese Versions

A: _____
B: _____
A: _____
B: _____
A: _____
B: _____
A: _____

B: _____
A: _____
B: _____
A: _____
B: _____
A: _____
B: _____
A: _____

Part 2 Intensive Reading

Text A

Better City, Better Life

Aiming to explore the 2010 Expo theme "Better City, Better Life", the Expo Forums, along with expositions and events, is one of the three major components of China's 2010 Shanghai World Exposition. Half of the world's population now lives in cities, and within two decades, nearly 60 percent of the world's people will be urban dwellers. The continuing urbanization process around the world has increasingly posed questions and challenges to all mankind.

ICT and Urban Development

The 21st century is a century of global urbanization, with cities playing an increasingly important role in a country or region. The ongoing ICT evolution reshapes and regroups traditional cities and transforms their social and economic bases enormously. Just like industrial development completely changed the spatial structure of cities in agricultural society, the progress of information and communication technology is the key element of the transformation of modern ones. Information and telecommunication technology as the choice of a future city, holds out the promises of a better life.

Science & Technology Innovation and Urban Future

Science and technology innovation has always supported and even accelerated the development of cities. The World Expos of the past were unique venues for countries to exhibit new scientific and technological achievements. Following the tradition of World Expos and promoting the concept of harmonious co-existence of people, cities and earth, Expo Shanghai 2010 will display the theme of "Better City, Better Life" by gathering the global know-how and wisdom.

Urban Responsibilities and Environmental Changes

Collective efforts of government, corporations and citizens are required to address major issues such as pollution, climate change, and energy conservation and emission reduction. In the face of environmental changes, all three are equally responsible for promoting a harmonious development pattern of environment, society, and economy. This could be achieved through new approaches to environmental management, and better mechanisms for regional and global cooperation. Environmental solutions can only be achieved through the joint commitment and actions of each community.

Harmonious City and Livable Life

Human beings live together and create cities for a better life. However, while people enjoy the advantages of urban development, they are also confronted with an increasing number of challenges. In an era of globalization and rapid urbanization, the quality of both the social and the built environment is essential to create livable life in the city. This forum will debate key issues such as: How to create a city that can satisfy the different needs of people's settled life and achieve the goal of "Better City, Better Life"? How to establish long-term mechanisms to secure harmonious urban development and therefore advance human society worldwide? All these might have been solved by seeking for feasible solutions through this Expo.

(447words)

Word Study

Words	Pronunciation	Versions
forum	['fɔ:rəm] n.	a public meeting or assembly for open discussion 论坛,讨论会,法庭
exposition	[ˌekspəu'ziʃən] n.	a collection of things (goods or works of art etc.) for public display 博览会,展览会
intellectual	[ˌintə'lektjuəl] adj.	of or associated with or requiring the use of the mind 智力的,聪明的,理智的
instrument	['instrumənt] n.	a device that requires skill for proper use 仪器,器械,工具,手段,乐器
urbanization	[ˌə:bənai'zeiʃən] n.	the social process whereby cities grow and societies become more urban 都市化
transform	[træns'fɔ:m] v.	change or alter in form, appearance, or nature 变换,改变,转化
spatial	['speiʃəl] adj.	pertaining to or involving or having the nature of space 空间的,存在于空间的,受空间条件限制的

accelerate	[ək'seləreit] v.	if a vehicle or someone who is driving it accelerates, it starts to go faster 加速,促进,增加
emission	[i'miʃən] n.	the act of emitting; causing to flow forth（光、热等的）发射,散发,喷射
livable	['livəbl] adj.	(of a house, etc.) fit to live in 适于居住的
mechanism	['mekənizəm] n.	the technical aspects of doing something 机制,技巧,原理,途径
essential	[i'senʃəl] adj.	extremely important and necessary in order to do something correctly or successfully 必要的,基本的
secure	[si'kjuə] v.	get by special effort 保护,招致,弄到

Proper Nouns

1. the Expo Forums 世博会论坛
2. World Exposition 世界博览会
3. ICT (Information and Communication Technology) 信息和通信技术

Useful Expressions

❶ aim to 目标在于……,以……为目标
　Aim to reduce road traffic accidents. 目的为减少道路交通事故。

❷ along with 连同……一起,与……一道,随同……一起
　Can you come along with us tomorrow? 你明天能来和我们一起去吗？

❸ be confronted with 面临,面对,对照
　The new system will be confronted with great difficulties at the start. 这种新的制度一开始将会面临很大困难。

❹ seek for 寻找,追求,探索
　Young people like to seek for success in life. 年轻人喜欢探索人生的成功之途。

Notes on Text A

❶ Aiming to explore the 2010 Expo theme "Better City, Better Life", the Expo Forums, along with expositions and events, is one of the three major components of China's 2010 Shanghai World Exposition.

　分析 aiming to... life 为现在分词短语,在句中作状语。

❷ Following the tradition of World Expos and promoting the concept of harmonious co-existence of people, cities and earth, Expo Shanghai 2010 will display the theme of "Better City, Better Life" by gathering the global know-how and wisdom.

　分析 following... and promoting... of people 都为现在分词短语,在句中作状语。

❸ This forum will debate key issues such as: How to create a city that can satisfy the different needs of people's settled life and achieve the goal of "Better City, Better Life"?

分析 key 在此处为形容词，表示"关键的"；satisfy ... needs 表示"满足……需求"。

Images connected with Internet 欣赏相关视频截图

城市人场馆主题:人的全面发展是城市可持续发展的前提
吉祥物:"海宝之家"
城市地球场馆主题:人类、城市、地球是共赢、共生的关系
吉祥物:"护林海宝"
城市生命场馆主题:城市如同一个生命活体，城市生命健康需要人类共同善待和呵护
吉祥物:"活力海宝"
城市足迹场馆主题:展示世界城市从起源走向现代文明的历程中，人与城市与环境之间互动发展的历史足迹
吉祥物:"海宝教授"
城市未来场馆主题:梦想引领人类城市的未来
吉祥物:城市未来馆以"太空海宝"作为视觉形象

会徽释义:

中国2010年上海世博会会徽，以中国汉字"世"字书法创意为形，"世"字图形寓意三人合臂相拥，状似美满幸福、相携同乐的家庭，也可抽象为"你、我、他"广义的人类，对美好和谐的生活追求，表达了世博会"理解、沟通、欢聚、合作"的理念，突显出中国2010年上海世博会以人为本的积极追求。

主体馆概况

上海世博会设有五个主题馆,其中城市人馆、城市生命馆和城市地球馆三个主题馆位于浦东B片区的主题馆建筑内。展馆外形设计从"折纸"的创意出发,屋顶则模仿了上海里弄"老虎窗"正面开、背面斜坡的特点,显示上海传统石库门建筑的文化魅力。

节能环保的现代建筑

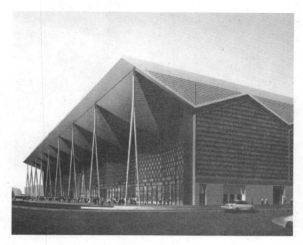

"老虎窗"式屋顶的外观有着一份上海里弄的怀念,既是近代上海建筑文化的象征,又结合了现代展馆的环保功能。它是世界上单体面积最大的太阳能屋面,太阳能板面积达3万多平方米,年发电量超过280万度,是四、五千户普通家庭一年的用电量,是目前太阳能覆盖率最高的单体建筑。

中国馆概况

展馆建筑外观以"东方之冠,鼎盛中华,天下粮仓,富庶百姓"的构思主题,表达中国文化的精神与气质。展馆的展示以"寻觅"为主线,带领参观者行走在"东方足迹"、"寻觅之旅"、"低碳未来"三个展区,在"寻觅"中发现并感悟城市发展中的中华智慧。展馆从当代切入,回顾中国三十多年来城市化的进程,凸显三十多年来中国城市化的规模和成就,回溯、探寻中国城市的底蕴和传统。随后,一条绵延的"智慧之旅"引导参观者走向未来,感悟立足于中华价值观和发展观的未来城市发展之路。

美国馆概况

展馆外墙设有"飞流而下"的瀑布媒体墙,顶部呈现生态环保的屋顶花园。馆内从"可持续发展"、"团队协作"、"健康生活"、"美国华人成就"四方面来演绎"拥抱挑战"的主题,展示美国的文化、价值观、创新精神和商业成就。

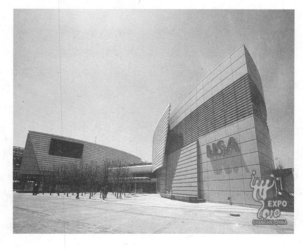

英国馆概况

英国馆的设计是一个没有屋顶的开放式公园,展区核心"种子圣殿"外部生长

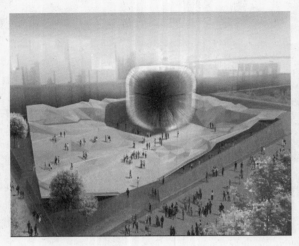

有六万余根向各个方向伸展的触须。白天，触须会像光纤那样传导光线来提供内部照明，营造出现代感和震撼力兼具的空间；夜间，触须内置的光源可照亮整个建筑，使其光彩夺目。展区中的"绿色城市"、"开放城市"、"种子圣殿"、"活力城市"和"开放公园"等几段参观旅程引导人们关注自然所扮演的角色，并思索如何利用自然来迎接城市面临的挑战。

沪上·生态家

"沪上·生态家"的原型是位于上海市闵行区的我国第一座生态示范楼。作为国内首座"零能耗"生态示范住宅，该建筑的一大优点是高效利用太阳能，屋顶上巨大的太阳能光热设备可为整幢楼提供能源。

外遮阳系统

外遮阳系统窗户外的百叶窗，落地玻璃外的卷帘门，还有阳台外的屈臂式遮阳篷，这些红色的装置叫做"外遮阳系统"，在炎热的夏天，这些装置能够随时阻挡阳光进入，起到隔热降温的作用。

地源热泵

墙上铺满的蓝色小管，就是利用地热能调节整幢建筑温度的地源热泵。通过管中流动的溶液将地下土壤中的温度带上来，使室内温度上升，就像中央空调一样。

Part 3 Extensive Reading

Text B

Konarka Power Plastic

Konarka Technology was founded in 2001 with such a vision: imagine a world free of carbon emissions; imagine a world where even the poorest most remote village has internet access and a light in every home; imagine a world where power plentiful, safe and truly green. Every day, they move closer to turning their vision into reality. The company's Konarka Power

Plastic Technology is lightweight, flexible, scalable and adaptable for use in a variety of commercial, industrial, government and consumer applications.

Konarka Power Plastic is a photovoltaic material that captures both indoor and outdoor light and converts it into direct current (DC) electrical energy. This energy can be used immediately, stored for later use, or converted to other forms. Power Plastic can be applied to a limitless number of potential applications—from microelectronics to portable power, remote power and building-integrated applications.

Konarka's Power Plastic offers several advances over other OPV technologies:

Power Plastic's tunable cell chemistry can absorb specific wavelengths of light—as well as broad spectrum. Unlike traditional OPV technologies, Konarka Power Plastic collects energy up to 70° off axis from sunrise to sunset outdoors—and anytime indoors.

Power Plastic is the only thin-film photovoltaic technology that uses all recyclable materials. Power Plastic applications fall into four main categories: microelectronics, portable power, remote power and building integrated applications (BIPV). Some of the applications currently under development by our manufacturing partners include:

· Portable battery chargers, for charging laptops, cell phones and lanterns

· Microelectronics, such as sensors, smart cards, remote starters

· Personal care devices, such as electric trimmers and toothbrushes

· Outdoor gear, such as tents and backpacks, which can power portable electronics and lighting

· Emergency power generators, which will enable police, military and emergency personnel to maintain vital communications

· Shade structures, which will power personal electronic devices anywhere the sun is shining

· Carport covers, which will trickle, charge an electric car

· Window shades and integrated window panels, which will capture and convert solar energy for both office and residential use

Solar energy represents the ultimate in green, renewable power. Konarka's break-through technology is transforming tomorrow's promise into an affordable manufacturing material for today.

With Power Plastic, energy independence will be very personal. Imagine a backpack that charges the laptop inside as you walk to work, a briefcase that powers your cell phone, a handbag that recharges your camera. Imagine an office window that powers your workplace computers and coffee makers. Imagine an umbrella that enables you to stay productive off the grid while you enjoy a day at the beach. Imagine a sun roof that powers your electronics while you drive, a car cover that recharges your electric car or a tent that turns on your reading light and warms up your sleeping bag after the sun goes down.

(464 words)

Word Study

Words	Pronunciation	Versions
access	['ækses] n.	a way of entering or reaching a place 通道,（使用或见到的）机会,权利
flexible	['fleksəbl] adj.	capable of being bent or flexed; pliable 灵活的,易适应的
scalable	[skeiləbl] adj.	if a piece of computer hardware or software is scalable, it continues to work well even if it is made bigger or connected to a larger number of other pieces of equipment 可伸缩的
photovoltaic	[ˌfəutəuvɔl'teiik] adj.	capable of producing a voltage when exposed to radiant energy, especially light 光电(池)的
portable	['pɔ:təbl] adj.	carried or moved with ease 手提式的,可移动的
integrate	['intigreit] v.	to make into a whole by bringing all parts together; unify, to join with something else 使结合(with),使并入(into)
spectrum	['spektrəm] n.	the distribution of energy emitted by a radiant source, arranged in order of wavelengths 光[波/能/质]谱,范围
charger	['tʃɑ:dʒə] n.	one that charges, such as an instrument that charges or replenishes storage batteries 充电器
trimmer	['trimə] n.	a device or machine, such as a lumber trimmer, that is used for trimming 修剪器,修整器
gear	[giə] n.	equipment, such as tools or clothing, used for a particular activity 设备,装置,机构工具
generator	['dʒenəreitə] n.	one that generates, especially a machine that converts mechanical energy into electrical energy 发电机
personnel	[ˌpə:sə'nel] n.	the development in a company that deals with employing and training people 人事部门
vital	['vaitl] adj.	most important, necessary to continued existence or effectiveness; essential 致命的,生死攸关的

trickle	['trikl] v.	to flow or fall in drops or in a thin stream; to move or proceed slowly or bit by bit 滴，缓慢移动
carport	['kɑ:pɔ:t] n.	an open-sided shelter for an automotive vehicle 简陋的汽车棚
ultimate	['ʌltimit] n.	the best, most advanced, great of its kind 最好（或先进、伟大等）的事物，精品

Proper Nouns

1. DC: a flow of electricity that moves in one direction only 直流电
2. AC: alternating current 交流电
3. BIPV: building integrated applications 建设综合应用；建设集成的应用程序
4. OPV: organic photovoltaic 有机光电伏电池
5. Photovoltaic Panel: (another name for a solar panel) 光电伏展板（或太阳能展板）
6. Konarka Power Plastic: 科纳卡光电塑料技术

Useful Expressions

❶ convert... into... 把……转换成……

He can convert defeat into victory. 他能反败为胜。

At what rate does the dollar convert into pound? 美元以什么汇率兑换成英镑？

❷ apply to 运用，适用

Would you apply that rule /principle to everyone? 这一规则[原则]能适用于每个人吗？

❸ enable... to do... 使……能做……

Training will enable you to find work. 培训将使你找到工作。

Computers enable people to get more information more quickly. 计算机使人们能够更快地获得更多的信息。

❹ warm up 热身，变暖，感到亲切，激动

Warm yourselves up before the game. 赛前活动一下。

I never bother to do warm-up exercises. 我向来懒得做热身运动。

Notes on Text B

❶ Konarka Technology was founded in 2001 with such a vision: imagine a world free of carbon emissions; imagine a world where even the poorest most remote village has internet access and a light in every home; imagine a world where power plentiful, safe and truly green.

分析 句中三个由动词原形imagine引出的祈使句作vision的同位语,其中where even the poorest most remote village has internet access and a light in every home 和 where power plentiful, safe and truly green 是由关系副词where引导的限制性定语从句,修饰先行词world。

❷ Konarka Power Plastic is a photovoltaic material that captures both indoor and outdoor light and converts it into direct current (DC) electrical energy.

分析 此句为主从复合句。is a photovoltaic material... and converts 是主句的并列谓语；that captures both indoor and outdoor light 是由关系代词that引导的限制性定语从句,修饰先行词material。

❸ Power Plastic is the only thin-film photovoltaic technology that uses all recyclable materials.

分析 此句为主从复合句。that uses all recyclable materials 是由关系代词that引导的限制性定语从句,修饰先行词technology。

❹ Imagine a sun roof that powers your electronics while you drive, a car cover that recharges your electric car or a tent that turns on your reading light and warms up your sleeping bag after the sun goes down.

分析 句中a sun roof, a car cover, or a tent 三个名词短语作宾语,后跟三个由关系代词that引导的限制性定语从句; while you drive 和 after the sun goes down 是状语从句。

Images connected with the Internet 欣赏相关视频截图

Photovoltaic energy in Missouri—Installation of photovoltaic panels on the roof of a warehouse in the town of Missouri (Figure 1)

Photovoltaic energy to the grid in Deltona, Florida—Solar energy can be thermal or photovoltaic. (Figure 2)

Photovoltaic energy in California (Figure 3)

Photovoltaic sale to the electric company in Santa Clarita, California Photovoltaic modules are the most visible part of a photovoltaic system, but, they are not the only component.

Solar power in Louisiana

Solar power plant of 35 kW on the roof of an industrial warehouse in Louisiana

Solar power in Georgia

Solar photovoltaic modules on flat factory roof for electricity sale

Photovoltaic energy in Missouri

Installation of photovoltaic panels on the roof of a warehouse in the town of Missouri

Part 4 Sentence Structures

Subjunctive Mood 虚拟语气

语气是一种动词形式,表示说话人对某一行为或事情的看法和态度。语气可分为下列三种:

(1) 陈述语气:表示说话人认为他所说的话是一个事实,如:

If he comes, he will bring his violin. 如果他来,他会带小提琴来的。

The volleyball match will be put off if it rains. 如果下雨,排球比赛将会延期举行。

(2) 祈使语气:表示说话人对对方的命令或请求,如:

Don't be late. 不要迟到。

Come along with me. 跟我来。

(3) 虚拟语气:虚拟语气是一种特殊的动词形式,用来表示说话人所说的话不是一个事实,而只是一种愿望、假设、怀疑、建议、猜测、可能或纯粹的空想,如:

If I were you, I would go there with him. 假如我是你的话,我就跟他一起去。

I wish he could come again. 但愿他能再来。

1. 虚拟语气在条件句中的用法

条件句分为两类,一类是真实条件句,一类是虚拟条件句。如果假设的情况是有可能发生的,就是真实条件句,如:

If I have time, I will come again. 如果有时间我会再来的。

如果是纯属假设的情况或发生的可能性不大的情况,则是虚拟条件句,如:

If he were here, it could be done. 如果他在这儿,这事就可以办。

因此,虚拟条件句主要是表示假设情况的。现将虚拟条件句的结构列表如下:

时间	条件从句	结果主句
与现在事实相反	If I (we, you, he, she, it, they)+动词过去式 (be 用过去式 were)	I(we)should+动词原形 you(he, she, it, they) would+动词原形
与过去事实相反	If I (we, you, he, she, it, they)+had done(been)	I(we)should+过去分词 you(he, she, it, they) would have +过去分词
将来渺茫的愿望	If I (we, you, he, she, it, they)+动词过去式 或 should do 或 were to do	I(we)should+动词原形 you(he, she, it, they) would+动词原形

虚拟语气的条件状语从句中的谓语如有 had, should, were 等助动词或实义动词 had, 从句连词 if 可以省略, 从句的助动词 had, should, were 或实义动词 had 置于句首, 用倒装语序, 如:

Were I you (=If I were you), I would go. 我要是你, 我就走。

Had they time (=If they had time), they would certainly come and help us. 假如他们有时间, 就一定会来帮助我们。

Should it be fine(=If it should be fine), we would go outing. 如果天气好的话, 我们就去郊游了。

Had it not been for my illness, I would have finished the work on time. 如果不是我生病了, 我就能按时完成这项工作。

2. 虚拟语气在宾语从句中的用法

(1) 用在动词 wish 后的宾语从句中, 表示无论怎样也不能实现的愿望, 如:

I wish I were a bird. 但愿我是一只小鸟。

We wish you could be a member of our group. 我们真希望你也是我们这个组的成员。

(2) 用在动词 ask, command, demand, request, insist, prefer, suggest, propose, order 等后面的宾语从句中, 表示请求, 建议, 命令等。其谓语用动词原形或"should+动词原形", 如:

She insisted that they give her a receipt. 她坚持要他们给她(开)一张发票。

I suggested that Mary (should) go to ask her teacher. 我建议玛丽去问问她的老师。

The doctor insisted that the patient (should) be X-rayed. 医生建议病人做X光透视。

3. 虚拟语气在主语从句中的用法

在 It is demanded / necessary / important/ natural/ a pity/ a shame + that...等结构后的主语从句中, 表示命令, 请求, 建议等, 其谓语用动词原形或"should+动词原形", 如:

It is necessary that he(should)act at once. 他必须立刻采取行动。

It is requested that all members (should) be present. 要求所有成员都到场。

It is a great pity that he (should) be so conceited. 真遗憾,他会这样自高自大。

4. 虚拟语气在表语从句中的用法

(1) 从句由 that 引导,主句的主语为表示建议,计划,命令等的名词,从句的谓语用动词原形或"should+动词原形",主句主语常用的名词有 aim, demand, desire, idea, motion, order, plan, proposal, suggestion, wish, advice 等,如:

My idea is that we (should) get more people to know the truth. 我认为我们应该让更多的人了解真相。

Our suggestion is that you (should) be the first to go. 我们建议你先走。

My advice is that we (should) send for a doctor at once. 我建议马上去请一个医生。

(2) 从句由 as if (as though) 引导,主句的谓语为联系动词,如:

You look as if you were ill. 你看上去好像生病了。

You look as if you had been ill. 你看上去好像生过病。

5. 虚拟语气在同位语从句中的用法

虚拟语气在同位语从句中主要表示命令,建议,计划等。从句的谓语用动词原形或"should+动词原形"。常与从句同位的名词有 advice, aim, demand, desire, idea, motion, order, plan, proposal, suggestion, wish 等,如:

He gave order that the guests (should) be hospitably entertained. 他下令要热情款待这些客人。

I make a proposal that we (should) hold a meeting next week. 我建议我们下周开会。

6. 虚拟语气在状语从句中的用法

表示方式的虚拟语气用 as if (as though) 引导,从句的谓语动词用过去式或过去完成式;表示目的的虚拟语气用 in order that 或 so that 引导,从句的谓语动词用 should (could/might) +动词原形;表示让步的虚拟语气用 even if (even though), though (although)引导,主句可以是陈述语气,从句的谓语用动词原形;表示原因的虚拟语气用 lest 引导,如:

Even if I were I his place, I would not go. 即使我站在他的立场,我也不想去。

Though the whole world should condemn her, I will still believe in her. 即使全世界都谴责她,我仍然相信她。

Whatever he might say, I'll stand my ground. 不管他说什么,我坚持我的立场。

He ran away lest he be seen. 他因为怕被别人看见而逃跑了。

She put her coat over the child lest he catch cold. 她把大衣盖在孩子身上,以免他着凉。

He took his umbrella with him lest it should rain. 他带上雨伞,以防下雨。

We were afraid lest he should get here too late. 我们唯恐他会迟到。

7. 虚拟语气在定语从句中的用法

虚拟语气用于定语从句,主要见于"It is (high) time that..."句型。从句谓语动词用过去式,that 常省略,如:

It is(high)time that we went to bed. 我们该睡觉了。

It is(high)time that you weren't here. 你该走了。

8. 虚拟语气的其他表现形式

I would rather you paid me now. 我宁愿你现在会付给我钱。(表示愿望)

If only I knew his name. 我要是知道他的名字该多好!(表示愿望)

If only the Babylon City hadn't disappeared. 当时巴比伦城没有消失就好了。(表示愿望)

You might try it again. 你不妨再试一试。(表示试探)

Did you wish to see me now? 你现在想见我吗?(表示试探)

Which seat might I take? 我坐哪个座位?(表示提问)

The foreign guests were to have attended our English evening. 外国朋友原本打算参加我们的英语晚会。(表示事与愿违)

I don't think he could be so careless. 我想他不会这样粗心的。(表示陈述)

Practice

a. *Choose the best answer to complete each sentence.*

1) You didn't let me drive. If we _____ in turn, you _____ so tired.

   A. drove; didn't   B. drove; wouldn't get

   C. were driving; wouldn't get   D. had driven; wouldn't have got

2) _____, we could not have finished the work on time.

   A. If it had not been for their help   B. Was it not for their help

   C. Should they offer to help us   D. If it is not for their help

3) _____ last Friday, he would have got to Paris.

   A. Was he leaving   B. If he is leaving

   C. Had he left   D. If he was leaving

4) If only I _____ how to operate a computer as you do!

   A. had known   B. would know

   C. should know   D. knew

5) When a pencil is partly in a glass of water, it looks as if it _____.

   A. breaks   B. has broken

   C. were broken   D. had been broken

6) The chairman requested that _____.

   A. the members studied the problem more carefully

   B. the problems were more carefully studied

    C. the problems could be studied with more care

    D. the members study the problem more carefully

7) It is important that he _____ called back immediately.

    A. was   B. should be   C. will be   D. were

8) We cannot imagine what our world _____ like without electric power.

    A. is   B. will be   C. would be   D. were

9) There was a smile on his face which suggested that he _____ happy to have given his life for his country.

    A. was   B. should be   C. would be   D. were

10) We all agreed to her suggestion that we _____ to the Great Wall for sightseeing.

    A. will go   B. go   C. shall go   D. should have gone

b. Reading comprehention

Photovoltaic energy in California

Photovoltaic sale to the electric company in Santa Clarita, California

Photovoltaic modules are the most visible part of a photovoltaic system, but, they are not the only component.

In the second image we can see some of the components of a photovoltaic installation for the sale of electricity to the grid. The left column is the CGP, the connection point between the electric company and the photovoltaic installation. The top box belongs to the photovoltaic seller and, the lower, which is open, to the electric company.

The remaining two columns consist of: ICP meters, transformers, fuses and a modem, all in the middle column and, differential switch, power on switch and intensity transformers, in the right column.

In the third image we can take a closer look at the connection point, called sectioning box, which belongs to the electric company and in the fourth image we see the two 32 kW invertors. The fifth image shows the inside of one of the invertors.

In fact, being in an urban environment is no problem for the use of solar photovoltaic energy.

Questions

1. This article is related with one of the pictures above. Can you find the picture?
2. This is an advertisement about how to install the photovoltaic panels. How can you find the images mentioned in this passage?
3. If you are interested in more information, please visit:

    http://www.sunandclimate.com/facilities/list/153-photovoltaic-energy-in-california.html

Part 5 Applying

1. Enjoy the Famous Sayings and Try to Recite Them.

A young idler, an old beggar
少壮不努力,老大徒伤悲。

Nature never deceives us; it is always us who deceive ourselves. ——Rousseau
大自然永远不会欺骗我们,欺骗我们的往往是我们自己。 ——鲁索

You can never plan the future by the past. ——Burke
永远也不能依照过去来计划将来。 ——佰克

Time is a versatile performer. It flies, marches on, heals all wounds, runs out and will tell. ——Jones
时间是个多才多艺的表演者。它能展翅飞翔,能阔步前进,能治愈创伤,能消逝而去,也能揭示真相。 ——琼斯

2. Creative Writing — on Building Materials

Make a speech about the materials of one of the buildings that you can find out in your daily life, and show some pictures related to these materials. Print the speech out in Paper A4 and send the speech to your English teacher's e-mail box.

Appendix 1 Keys to Exercises 练习答案

Lesson 1 The Ancient Chinese Architecture

a. 时态选择题 1–5 答案：D B D B B

解析：

1. 译文：我不知道是否会下雨，如果下雨，我将呆在家里。
   在时间或条件状语从句中，常用一般现在时表示将来的动作或状态。
2. 译文：她刚放下听筒，电话就又响了。
   Hardly/Scarcely... when... 为固定结构，表示"刚……就……"，主句要用过去完成时，从句用一般过去时。
3. 译文：到期末考试时，他将听完五个讲座。
   从引出的时间状语可知，主句的谓语表示将来某时刻之前完成的动作。所以用将来完成时。
4. 译文：火车刚走，你要是早来五分钟就好了。
   在 if 引导的感叹句中，谓语动词用虚拟式，前句是对过去的假设，因此要用过去完成时。
5. 译文：明天的这个时候，他们正在爬山。
   从时间状语看，谓语是某一时刻发生的动作。所以用将来进行时。

b. 翻译

1. 下一班列车的发车时间是今天下午三点钟。
2. 你现在是我们学院的理科生吗？
3. 我要等你三个小时，你才做完报告。
4. 在上大学前，我已经学了 5000 个单词。
5. 到上月底，我们已经按照协议完成了这一工程。
6. 我们将送她一个手工的玻璃飞机作为生日礼物。
7. 我已经在这儿工作了三年。
8. 我正要给他打电话，他的信就到了。
9. 他习惯冬天游泳。
10. 希望明年你回来时，他的健康状况已经有所改善。

c. 语态选择题 1–5 答案：C D C B D

解析：

1. 译文："今天很冷，是吗？""是的，河水已结冰。"
   本题考查两点：第一，被动语态；第二，一般现在时的用法。
2. 译文：他上大学时很讨厌数学。
   本题考查要点：一般过去时的被动语态。be bored with, "对……厌倦"；was bored "感到厌倦"，后面不能跟名词。
3. 译文：我叔叔刚当选为公司经理。
   本题主要考查现在完成时的被动语态。

4. 译文：每年都花费大量资金粉刷钢构桥梁。
   考查情态动词 have to 后接被动语态。

5. 译文：申请这个职位的应聘者正在接受面试。
   本题主要考查现在进行时的被动语态。

Lesson 2 The Trend for Constructions in China

选择题1-10答案：C D A D C A A A D B

解析：

1. 译文：他们几乎不在家吃午饭，不是吗？
   seldom 具有否定含义，反意疑问句用肯定形式；本句中 have lunch 的 have 是实意动词，而非助动词，所以选C。

2. 译文：这位老太太记不清把钥匙放哪里了。
   该特殊疑问句非直接引语，作宾语是宾语从句，应用陈述句语序，并不需要 that 作连接词。

3. 译文：天真热啊，不是吗？
   本句是个省略的感叹句，What a hot day (it is), _____?

4. 译文：这就是警察在找的东西吗？
   本句是 what 引导的宾语从句。

5. 译文：我女儿想了解一下去东京度假的信息，不是吗？
   陈述部分的主语是 My daughter，疑问部分中的主语用 she 替代。

6. 译文：你永远不可能知道下一步该怎么做，不是吗？
   never 具有否定含义，反意疑问句用肯定形式。

7. 译文：不要在这里吸烟，可以吗？
   此句为否定形式的祈使句，所以选A。

8. 译文：我们都没有被要求出示票据，不是吗？
   Neither...nor... 意为"两人都不"，因此，反意疑问句用肯定形式，人称代词用复数。

9. 译文：皮特喜欢踢足球但不喜欢弹钢琴，不是吗？
   就近原则。dislikes 为否定意义，所以选D。

10. 译文：木匠们曾经居住在加拿大，是吗？
    used to do sth. 过去式，肯定形式，所以选B。

Lesson 3 The European Art of Building

1. How beautiful the flower looks in the garden!
2. What an exciting day we have had!
3. How foolish I was to think like that!
4. What lovely weather it is!
5. How proud we are of our great motherland!
6. What a fine voice he has!

Lesson 4 The American Architectural Style.

选择题1-10答案：D C C D A B C D B D

解析：

1. 译文：你是想休息一下还是想继续工作？
   此题考查并列句中连词的使用。

2. 译文：要是再早五分钟的话，我们本可以赶上最后一班火车的。
   此题考查并列句中连词的使用。

3. 译文：尽管已经告诉他多次了，但他还是无法理解。
   此题考查并列句中连词的使用，A项和D项中都用了现在分词，B项中"though"和"but"不能连用。

4. 译文：汤姆的妈妈一直都告诉儿子要好好学习，但这毫无用处。
   此句子为一个由连词"but"连接的并列句，因此不能用关系代词"which"，空格中需要的是"妈妈告诉儿子要好好学习"这件事情，因此用it指代。

5. 译文：妈妈病得很厉害，因此玛丽很难过。
   此句为一个因果关系的并列句，B和C两项中都是用了现在分词。D项中Because和so不搭配不能连用。

6. 译文："加牛奶和白糖吧。""只要牛奶吧，但以前我是喜欢加糖的。"

7. 译文：他们很少在家吃午饭，不是吗？
   此题考查反意疑问句，反意疑问句的陈述部分带有little, few, never, hardly, seldom等否定意义的词时，问语部分用肯定式。

8. 译文：无论是谁知道真相都肯定会告诉你的。
   此题考查主语从句，根据句意应选择"whoever"，表示"无论是谁"。

9. 译文：詹尼佛阿姨尽管已经四十岁了，但看起来只有二十岁。
   此题中"she were only 20 years old"用了虚拟语气，故选择B "as if"。

10. 译文：有很多理由能使我振作起来。
    此题考查定语从句，先行词在句中作主语，故使用关系代词"that"。

Lesson 5 The Architectural Art in Asia

选择题1-5答案：C A B D A

解析：

1. 译文：电脑只能做有指令的事情。
   what引导宾语从句，what you have instructed it to do作动词do的宾语。

2. 译文：—上周我开车去珠海参加了航空展。
   　　　—那就是你有几天没来的原因吗？
   why引导表语从句。

3. 译文：我讨厌人们嘴里吃着东西时讲话。
   that引导同位语从句。

4. 译文：英语正在成为一门国际语言，这已成为事实。
   it作形式主语，真正的主语是English is being accepted as an international language.

5. 译文:我仍然记得这曾经是个静谧的村庄。
   when引导宾语从句。

Lesson 6 The Classical Masterpiece in Ancient China
选择题1-10答案:A C D C A C A C A A
解析:
1. 译文:我最感兴趣的地方是少年宫。
   关系词在从句中作主语,由此确定是关系代词,排除where和in which;what不充当关系词。
2. 译文:你认识跟我说话的那个人吗?
   根据句意,从句表示与之说话的人,关系词在从句中作介词宾语,作宾语的关系词可以省略。由此选C。
3. 译文:这就是他们上个月住过的旅馆。
   先行词hotel,关系词在从句中作地点状语,可以用where引导,也可以用at which引导,不可用at that引导,所以选D。
4. 译文:你知道中国共产党是哪一年成立的吗?
   先行词year,关系词在从句中作时间状语,可以用when也可以用in which,所以选C。
5. 译文:那是我永远也不会忘记的一天。
   先行词是day,关系词在从句中作宾语,只能用which或者that,所以选A。
6. 译文:我们下周要去的那家工厂离这里不远。
   先行词factory,关系词在从句中作宾语,只能用which或者that,所以选C。
7. 译文:我们曾经工作过的那家工厂从那之后发生了很大变化。
   先行词factory,关系词在从句中作地点状语,可以用in which,which... in,that ... in,也可以用where。
8. 译文:这是今年放映过的最好的电影之一。
   根据句意,应该是被动语态的现在完成时态,所以选C。
9. 译文:你能把那天你说的那本书借给我吗?
   先行词book,关系词在从句中作谈论talk about的宾语,关系词可以用that,也可以用which,但that前面不可以用介词,所以选A。
10. 译文:他用来写字的笔是我的。
    先行词pen,这是由"介词+which"引导的定语从句。write... with指"用……写字",所以选A。

Lesson 7 The Ancient Architecture in the World
选择题1-10答案:B A C C A B B A B D
解析:
1. 译文:除非你16岁了,否则你不能考取驾驶证。
   unless(除非)引导条件状语从句,常用的引导词还有if(如果), so long(只要), provided that(假如,假设), in case that(万一), lest(以免,唯恐,免得)等。

2. 译文:这个年轻人上个月失业了,但是没过多久就在我公司找到了一个新的职位。

   before引导的时间状语从句。用于句型"it was+时间段+ before..."表示"过了(多长时间)才……"。其否定形式"it was not +时间段+ before..."意为"不久就……","没过(多长时间)就……"。本题的意思是"不久就","没过(多长时间)就……"。

   例:It was some time before we realized the truth. 过了一段时间才意识到事情的真相。

3. 译文:如果你有任何问题或需求,请在周末下午5点之后和经理联系。

   本题中if引导条件状语从句。引导条件状语从句(真实条件句)的引导词有if(如果),unless(除非),so long as(只要), provided that(假如,假设),in case that(万一)等。

4. 译文:每次苏珊到达高楼顶部,就感到很害怕。

   every time引导的时间状语从句。

5. 译文:每次他外出都要戴上太阳镜,以便没人能认出他。

   so that引导目的状语从句。引导目的状语从句常用的引导词有so that (为了,以便), in order that (为了,以便), for fear that (唯恐,害怕), lest (以免), in case (以防,以免)。

6. 译文:我假期住的那个旅馆条件很差。

   where在这里引导定语从句。先行词the hotel在从句中做状语,"I stayed in the hotel." 所以选B。

7. 译文:直到去年李雷在美国度假时才见到那位著名的美国教授。

   not...until...引导时间状语从句,"直到……才……"。

8. 译文:她昨天没有去电影院,因为她要完成学期论文。

   as引导原因状语从句。引导原因状语从句的从属连词有because(因为),as(由于),since(既然),now that(既然),in that(以为)等。because从属连词,语气最强,可放在句首,也可放在句尾。回答why提出的问题时必须用because;as语气不如because和since强烈,只是附带说明;since语气比because轻,但比as重,很少用于口语。

9. 译文:他们刚下火车,火车就开动了。

   no sooner... than, hardly (scarcely, barely) ... when 表示"刚……就……",一般前面用过去完成时,后面用一般过去时,如no sooner或hardly在句首,要用倒装语序。

10. 译文:对于强壮的人来说,要意识到无论他多么强壮,都不是最强壮的,这一点是很重要的。however(无论多么)引导让步状语从句。常用的引导词有even if(即使), though/although(虽然), even though(即使), while(尽管), no matter what/how/which(不管什么/怎样/哪个), whatever(无论什么), whichever(无论/哪个), whether... or (不论……否), as(虽然) 等。

Lesson 8 Ecology and Architecture

a. 选择题1–10答案:D A C D C D B C B B

解析:

1. 译文:你当时没有让我开车。如果我们轮流开车,你就不会那么疲惫了。

   本题考查虚拟语气。主句、从句都是对过去的假设,所用的动词形式应分别为had + p.p.或did和would + have + p.p.。

2. 译文:如果没有他们的帮助,我们就不能按时完成工作。

   本题考查虚拟语气。主句、从句都是对过去的假设,所用的动词形式应分别为had + p.p.或did和would + have + p.p.。

3. 译文：如果他上星期五走，就已经到巴黎了。

   本题考察倒装和虚拟。主句 would have done 表示是与过去事实相反的虚拟语气，条件从句中对过去的虚拟要用过去完成时态。

4. 译文：要是我像你一样懂得如何操作电脑就好了。

   本题考查虚拟语气。only +句子(过去时/过去完成时)表示"要是……就好了"。If only 后面的句子如果是对现在或将来情况的虚拟，用过去时 did。

5. 译文：当一支铅笔的一部分置于水中时，它看上去像是被折断了。

   本题考查被动语态和虚拟语气。as if 引导的状语从句中，用过去时表示与现在事实不符，用过去完成式表示与过去事实不符，本句是一般现在时。

6. 译文：主席要求会员更加仔细地研究这个问题。

   本题考查虚拟语气。当句中出现表示建议、命令、要求的词，如 demand, request, require, order, advise, suggest 等，其从句需用虚拟语气，谓语动词用"should+动词原形"，通常 should 可省略。

7. 译文：重要的是马上叫他回来。

   本题考查虚拟语气和被动语态。important 用在 It is + a. + that 句型中，其后的句子要用虚拟语气，谓语动词用 should 加动词原形，且 should 可以省略。

8. 译文：我们难以想象如果没有电力，我们的世界将会怎样。

   本题考查虚拟语气。表示与将来的事实相反，或对将来的推测。

9. 译文：他脸上的笑容表明他很高兴能为祖国献身。

   本题考查虚拟语气。在动词 suggestion, order, demand, proposal, request, command, insist 等后的宾语从句中，用虚拟语气(即 should+动词原形或只用动词原形)来表示愿望、建议、命令、请求等。注意：如 suggest, insist 不表示"建议"或"坚持"要某人做某事时，即它们用于其本意"暗示、表明"、"坚持认为"时，宾语从句用陈述语气。

10. 译文：我们都同意她的建议，去长城观光。

    本题考查虚拟语气。在名词 suggestion, order, demand, proposal, request 等后的同位语从句中，用虚拟语气(即 should+动词原形或只用动词原形)来表示愿望、建议、命令、请求等具体内容。

b. 阅读理解答案：

1. Photovoltaic energy in California (Figure 3) on page 118 shows this.
2. By the Internet.

Appendix 2 Chinese Versions for Text A & B 参考译文

第一课 中国古代建筑

Listening and Speaking 中国建筑哲学理念

A: 据我所知,你目前在学习建筑吧?
B: 是的,我非常喜欢。
A: 你最喜欢什么样的建筑?
B: 各种风格的我都喜欢,但是最喜欢的还是亚洲大陆建筑,尤其是中国的建筑。
A: 真的吗? 你知道中国的主要建筑理念是什么吗?
B: 因为注重平衡,中国的很多建筑结构都是对称的。同时,中国建筑还喜欢在横梁和过梁上使用曲线和雕刻。非常注重细节。
A: 我觉得很多国家的建筑也都采用了中国的一些建筑理念。我在很多国家都见过带立柱的亭子。
B: 中国建筑的精彩之处就在于它在发扬古代传统的同时又吸纳了一些新的技巧和现代流行趋势。你能认真学习和潜心研究建筑风格真是一大幸事。

Text A 中国古代建筑简介

中国位于亚洲大陆的东部,陆地面积约960万平方公里,人口占世界总人口的1/5,拥有56个民族和5000多年的历史。中国悠久的历史创造了灿烂的古代文明,而独特杰出的中国传统建筑便是其重要组成部分。

建筑理念

所有中国古代建筑都是根据道教和其他各学派传统理念而设计的。第一个建造理念是建筑物应长而低而不是短而高,看起来好像要把人围起来一样。屋顶由木梁支撑而不是由墙体支撑。屋顶看起来像漂浮在地面上。第二个理念是对称。建筑物两边的结构相同且对称,这恰好是遵循了道教所推崇的平衡原则。即使在公元前1500年前的商朝建筑也都是相当对称的,有着上翘瓦屋面和长排的立柱。

建筑特点

中国古代建筑总体上说以木结构为主,有着独特迷人的外表。与主要以砖石结构的世界其他国家建筑不同。木质的柱子,横梁和托梁构成房屋的框架。墙壁起分隔房间的作用而不是支撑作用,这是中国古代建筑的独特之处。正如中国民间俗语所说"墙倒屋不塌"。中国古代建筑的特点之一是大胆使用色彩。这个特点是和中国建筑的木结构体系分不开的。因为木料耐久性差,需要采取专业的防腐措施,后来就发展成中国自己的建筑彩绘艺术。琉璃的屋顶、窗户以及其精湛的设计,还有漂亮的用花朵图案贴花的木柱子都反映了古代工匠们高超的技艺和丰富的想象力。

建筑与文化

建筑与文化是紧密相连不可分割的,在某种意义上可以说建筑是文化的载体。中国古代建筑风格丰富多变,例如庙宇、宫殿、祭坛、楼阁、官邸和民宅。这些不同类别的建筑都体现了中国古代的思想——天人合一理念。

中国古代建筑文化的典型代表是风水和牌坊。

风水:中国传统理论尤其强调建筑要以《易经》文化为基础,强调人与自然的和谐统一。

牌坊:(也叫牌楼)它在世界上是独一无二的,它是封建社会为表彰功勋和忠孝节义所立的建筑物,它是中国封建社会文化的产物。

今天，以其传统风格为基础的中国建筑吸收了外国建筑文化，并遵循我们时代的要求，采用日益更新的建筑技术不断开拓进取。

Text B 长城

中国长城是世界上最伟大的奇迹之一，1987年被联合国教科文组织确定为世界文化遗产。长城西起甘肃省的嘉峪关东到河北省的山海关，宛如一条巨龙，绵延经过沙漠、草地、高山和平原，总长约为6700公里。作为世界八大奇迹之一，长城成为中华民族和中国文化的象征。

长城由山西省作为分界线分为东西两部分。西部长城是夯土结构，平均宽度为5.3米。东部长城内层也是夯土结构，外层则由砖石加固。长城最壮观和保存最完好的部分是位于北京附近的八达岭长城和慕田峪长城。

长城高约25英尺(7.5米)，宽30英尺(9米)。城墙顶部是砖石铺成的路，可供10个士兵并排行走。墙的顶部有防御土墙、射洞和擂石孔供弓箭手使用，还有顺着矮墙建造的带有滴水嘴的排水沟。大约每间隔400米就建有两层的烽火台。烽火台的上层用来观望敌情，下层用来提供食宿、供应马匹粮秣等服务。八达岭长城最高处的烽火台建在陡峭的山崖之上，就像通向天空的梯子。登上这里的长城，可以居高临下，尽览崇山峻岭的壮丽景色。长城随山脉绵延起伏，直到最终消失于雾霭中。

沿长城地势险要之地共建有14个关隘，最重要的当属山海关和嘉峪关。众所周知的天下第一关"关山海"坐落于两个陡峭的悬崖之间，是连接中国北方和东北方的咽喉之地。因此这里成为历来的兵家必争之地，许多著名战役也发生在这里。当年正是明将吴三桂打开山海关引清军入关，打败了李自成领导的农民起义军，导致了明灭清兴(1644–1911)。

长城是最伟大的历史文化建筑，这也是它具有世界性吸引力的魅力所在。

第二课 中国建筑业的发展趋势

Listening and Speaking 长城

A: 哇,那长城太壮观了!

B: 是啊。大约有25米高,15米宽呢。

A: 为什么要建造这么宏大的长城呢?

B: 它主要是用来保卫城市,防御外敌入侵。敌人要想入城就必须用大锤把城墙撞开。看到那些矮护墙了吗?当年守卫的士兵们就是从这里观察敌营,对敌兵一览无余。

A: 天啊,这需要多少兵力才能攻破长城,让城内的军队投降。尽管它是用来防御的,但其结构却也很吸引人。

Text A 中国民居的特点

民居是中国历史上最早的建筑形式。这些住宅通常不受什么限制,形式多样,并符合当地条件。他们具有强烈的民族风格,是体现本地特色的中国建筑艺术。中国幅员辽阔、历史悠久,有着不同的自然和人文环境。因此,各地的民居风格也各不相同。主要有北京的四合院,陕西黄土高原的窑洞,湖南西部建于陡坡或水上的吊脚楼,安徽南部的徽州民居和福建客家土楼等,这些古老的建筑都独具特色。

首先,中国民居的美在于人与居住空间的和谐,建筑形式适应人们的生活需要;同时,建筑材料均产于当地,民居色彩简单纹理优雅。此外,建筑的布局合理,简洁实用。

就中国民居的装饰而言,建筑主体一般采用雕刻和彩绘,辅以极少的其他装饰元素。主要的装饰技术包括砖雕、木雕和石雕,用于装饰大门、门窗的框格、影壁墙、屋顶和防火墙这些重要、突出之处。长江以南和安徽南部的民居装饰是最雅致、精美的,尤其是安徽南部三种高超的房屋装饰技术——砖雕,木雕和石雕,均以无与伦比的技术而闻名。

不同地区的民居具有鲜明的地方风格和民族特点。装饰风格也会因业主的经济地位和教育背景不同而不同。知识分子房屋的装饰风格和品位与大富豪的品位大相径庭。一般来说,民居都古朴、典雅。许多房屋的大门两边都挂有对联。楹联的内容、书法既生动又质朴。厅堂中挂的对联行文缜密,摆放的家具豪华,但很实用。装饰雕刻大多颜色静谧,给人以优雅的感觉。

民居不仅用于居住,而且反映了不同的风俗,诉说着不同的生活故事。品味民居的艺术就是体验多元化的中国民族文化和各民族的生活历史。

Text B 中国北京的现代建筑

中国给全世界带来了长城,宝塔屋顶,屏风,接榫和许多园林创新建筑。然而,那都是几百年前的事了。近代中国展现在世人面前的大都是混凝土建筑物。

继20世纪90年代建设的蓬勃发展后不久,内地城市的建设再次处于静止状态。中国正决心开创社会主义现代化建设新时代。

在著名的毛泽东画像下,天安门广场的大多数建筑物都很宏伟壮观,有着硬朗的线条和大型的圆柱。他们具有明显的时代性,即革命后共产主义中国的特征。

但在不远处的一座引人注目的新建筑,即新的国家大剧院,却与广场上其他建筑有着根本的不同。新的国家大剧院座落在水中,建筑屋面呈半椭圆型,由钛金属板和玻璃墙体组成。它的入口处在地下。该建筑可能是北京过去几年里最有争议的一个。

"目前此处的这个现代化建筑,让人看起来很不舒服,事实确是如此,"它的设计者,法国建筑师保罗安德鲁说。

除了安德鲁的建筑,中国政府还把其他一些大项目,授予了外国建筑师。鸟巢体育场是由瑞士建筑师设计的,中国国家电视台——中央电视台新总部是由荷兰人设计的。

一些钢铁框架构成了北京国家体育馆的圆顶,这里也是举办2008年北京夏季奥运会的奥林匹克体育馆。然而,北京的许多年轻人对这种光怪陆离并不特别感兴趣。

"我不喜欢它,"罗晴,32岁,她看着中央电视塔从她的办公室的街对面升起来。"我不能接受这种外表摩登的建筑。我还是喜欢传统的中国建筑。"

罗晴的同事孙鹏,也32岁,赞同她的观点。

"我认为北京应该建造能反映城市历史和文化的建筑物,特别是一些重要的地标建筑,如中央电视塔,"孙鹏说。

"我相信,文化是重要的,"建筑师安德鲁,站在被他称为"湖中之文化岛"的蛋形中国大剧院前说。

但是,"似乎它在许多方面都与北京的背景不怎么相称"研究大陆建筑史的香港中文大学教授杰弗里科迪说。"安德鲁觉得他很委屈,但大多数人都认为他没有把握好新潮与创新之间的分寸,不能与老百姓产生共鸣。"

马国馨曾于20世纪80年代早期在日本丹下健三建筑室做访问学者,他出于对传统的尊重认为,中国的准则——例如天人合一的理念——能够把西方的理念转变成具有中国特色的东西。他并不担心中国的美学会在现代化进程中受到排挤,他相信目前中国大胆聘用外国建筑师来搞城市建筑规划设计会给中国带来新技术,使中西方建筑文化产生碰撞,最终共同孕育出当代中国建筑风格。

"我们的传统文化是如此有魅力,它将永远传承下去"他说。

第三课 欧洲建筑风格

Listening and Speaking 合作施工

A：约翰逊先生，你好，我叫史蒂芬，我是这个工程的建筑师之一。很高兴见到你。

B：谢谢。叫我迈克就行了。我很高兴工程启动了。我雇佣你们公司是因为你们公司在市区工程干得很出色。

A：好的，有这么好的声誉我也很开心。您是不是想在花园里建一座宝塔？

B：对，我想让它成为一个真正的景观，希望建成后宝塔能和周围的景色产生共鸣。

A：我们一起来看一下图纸吧。看起来地基和支柱都是用混凝土，对不对？

B：是的，我想让它的结构更坚固。还有什么比不结实的建筑更令人感到不安的呢。而且我不希望再去担心白蚁的入侵。你觉得屋顶设计得怎样？

A：太漂亮了。你使用的全是最先进的材料，这很耐用啊。

B：是啊，这正合我意。我希望它看起来即壮观且具有皇家风范，但又不至于让整个花园逊色。

A：很好。我们马上就开工。

Text A 中世纪的欧洲建筑

在中世纪，许多不同的建筑风格得到了发展。那些试图模仿罗马穹隆与拱券的教堂现在被称为"罗马风"建筑。后来使用了尖券和更富于装饰性的花饰窗格的教堂被称为"哥特式"建筑。哥特式建筑又进一步细分成不同的地方风格。在英格兰，我们发现有早期英国哥特式、盛饰式与垂直式等风格。在法国，则存在盛期哥特、火焰纹哥特以及辐射式哥特风格。这些地方哥特式建筑风格的名称大多源自描述建筑效果的概念，或是花饰窗格的式样。"罗马风"也将我们引回到建筑形式，而这些建筑形式又让人想起了古罗马的建筑。在英格兰以及法国北部，"罗马风"建筑又被称为"诺曼式"建筑，这是以他们的诺曼底领主来命名的，因而是对同一建筑风格所命名的"政治性"的称谓。谕旨罗马大帝的子民。同样，"哥特式"这个名称也是政治性的，因为这一名称与一个族群，也就是消灭罗马的"哥特人"有关。这个名称并没有告诉我们任何有关建筑的信息，但它却告诉了我们在出现这个名称的17世纪，人们是如何看待这一类建筑的，虽然教堂建造者早已不再使用这种方式建造教堂了。

中世纪最雄伟的标志性哥特式建筑是法国大教堂，像圣·艾蒂安大教堂。位于法国布尔日的圣·艾蒂安大教堂，在所有中世纪的教堂中，可以说非常完美地诠释了光笼这一建筑理念。教堂的西端相当坚实，5个大小不等的入口由几百个小型雕塑包围着。在这些入口上方是两座不同的塔楼。然而，教堂的其余部分却给人一种经过精确复制的印象——以一种标准的开间形式，沿教堂的长度方向连续排列不间断，而且偏差很小，以使其能够在建筑物西端形成半圆。中厅高耸，有39米（125英尺）之高，中厅的两侧是两个侧廊。阳光透过绘有《圣经》故事的彩色玻璃照进来，使得教堂内部十分敞亮。从外观看，用以支持建筑物主体的扶壁立柱很是显眼，看起来就像是一排气势恢弘的支柱：正是这些扶壁立柱使得教堂内部给人一种轻盈飘浮的幻觉效果。

Text B 新古典主义建筑

新古典主义建筑作为一种建筑风格，是对洛可可式建筑风格室内装饰的反叛，以及后期巴洛克中一些仿古典特征的副产物。它始于18世纪中叶的新古典主义运动。这场被称为新古典主义的运动还受到在希腊的考古发掘中所获得的新知识的滋养。这些考古发掘使人们更加清晰地了解了古希腊的建筑形式。

采用这种建筑风格的主要是银行、博物馆、剧院等公共建筑和一些纪念性建筑,而对一般的住宅、教堂及学校影响不大。

法国在18世纪末、19世纪初是欧洲新古典建筑活动的中心。法国大革命时在巴黎兴建的万神庙是典型的古典式建筑。拿破仑时代在巴黎兴建了许多纪念性建筑,其中雄师凯旋门就是古罗马建筑式样的翻版之一。雄师凯旋门是一座纪念碑,位于法国巴黎戴高乐广场中央,这个广场又被称为星形广场。它位于香榭丽舍大街的西端。凯旋门是为了纪念那些为法国英勇奋战的士兵,特别是在拿破仑战争中奋战的勇士们。凯旋门高49.5米(153英尺),宽45米(150英尺),厚22米(72英尺)。是世界上现存第二大凯旋门。它的设计灵感来自于罗马的提图斯凯旋门。

随后到了18世纪末,一些与前期古典主义形成竞争之势的古典风格流行开来,这些风格以对希腊与罗马建筑的不同理解为基础,从主张造型严谨的多立克神庙那种纯正的简单性,到亚当兄弟那些高度装饰化的作品,不一而足。同时,还有一种日益增强的复兴中世纪建筑的复古情绪,后来演化成了19世纪中叶的哥特式复兴建筑风格;各种异域风情的尝试也层出不穷,如皇家穹顶宫殿就是一例。自此开始,折中主义风潮开始兴起,并在某些时期表现得尤为明显,这说明了这样一个事实:那些能在建筑上花费巨资的人的审美情趣已不再统一了。

第四课 美国建筑风格

Listening and Speaking 金字塔和寺院

A: 欢迎归来！您的南美之旅怎么样？有没有看到一些宏伟的建筑？

B: 旅行很棒。我看到了很多建筑。我们大部分时间都停留在玛雅地区。

A: 我听说你在那儿待了没有多久就感受到了当地的建筑特色。

B: 那里的主要建筑是金字塔和寺庙。这些石头建筑绝对是一流的。

A: 能完成这样的建筑,玛雅人的文明程度很高啊！

B: 让我感到最有趣的是他们的建筑方式质朴,但同时又相当精美。金字塔非常紧凑,主要是用石头建成的。

A: 希望您能慢慢品味这次旅行,真正去回味它。难得您能有机会去参观这样一个具有民族特色的美景。我真为您高兴。

Text A 美国建筑业的发展状况

纵观美国历史,其建筑经历过各种各样不同的风格。美国建筑具有地区差异性,并且受许多外部因素影响,因此可以说是兼收并蓄的,这种特点在这样一个多元文化国家是不足为奇的。

外观样式和相关建筑形式以及建筑平面图在某种程度上是一种文化品位和价值观的产物,反映了某一特定的地点,时间,和人口状况。风格有点类似于服装潮流,它随着时间的推移来来回回,有时也会再回到原先的起点。即使在全国时尚和文化理念处于低潮时期,某些建筑风格也会风行几十年甚至更长的时间,并且往往显示出明显的区域特征。

到了19世纪中晚期的维多利亚时代,多种建筑风格在美国同时开始流行并且迎来了历史学家称之为"折中时代"的建筑时期,当时美国人倾向于现代或复兴的建筑风格。这种所谓的"时期风格"和早期现代主义共存的魅力一直持续到大萧条时期,并且有增无减。之后,其实在1929年至1945年间很少有建筑动工。

直到二战结束,美国才迎来了又一次国家建设热潮。当时,趋于流行的建筑是利用汽车出行的郊区住所、现代化公寓住房以及商务办公大厦。从20世纪40年代到80年代,盛行美国的是现代功能主义,人们对传统历史建筑风格较为反感。由于形式依赖于功能，人们熟悉的"玻璃盒子"似的办公大楼和无处不在的平房仍然是这种反风格时代的主要象征。

但是，这种形式在20世纪70年代发生了变化，美国人开始抵制现代建筑，于是历史风格逐渐再次流行，恰逢此时迎来了蓬勃发展的历史保护运动。用带有殖民地复兴元素的风格装饰现代派的平房，于是到20世纪90年代，一种模糊的"后现代建筑时代"全面盛行开来。后现代建筑的普遍特点是对一些毫不相关的历史风格的夸张运用，或者是对一些古旧建筑的复制模仿。后现代历史主义的兴起恰逢以非传统或新都市主义著称的振兴城镇建设再次流行。这种建筑顶端设计成高耸的阁楼和公寓式样，而且具有多种用途。于是，很快建成或在建的新型传统社区就达到数以百计，这种社区的设计注重行人步行，公共交通设施，功能多样化，社区可居住性，以及公共空间绿化带，而且价格低廉。

美国的建筑风格依然在不断变化和发展。我们不禁要问，美国下一个重要的文化情趣是什么，建筑环境又将如何反映这一情趣？

Text B 中美建筑艺术家

世界各地历史上有许多建筑师，他们为世界建筑的发展做出了重要贡献。以下列举一些中美著名建筑师，他们都拥有很多建筑作品并因此而享誉世界。

梁思成（1901-1972）是中国最有名的建筑师之一。他是晚清著名思想家梁启超之子。梁思成著有中国第一部现代建筑史，于1928年创办东北大学建筑系，于1946年创办清华大学建筑系。他曾作为中国代表参与设计位于纽约的联合国总部大厦，被称为"中国现代建筑之父"。普林斯顿大学1947年授予他荣誉博士学位，并对他进行了高度赞赏"作为一名建筑史老师，他也一直是一位创造性的建筑大师，建筑历史研究和中国建筑规划探索的先驱，并在恢复和保护其祖国的珍贵古迹的运动中发挥了领导作用"。

贝聿铭（1917-），人们亲切地称他为 I. M. Pei，他是一位华裔美国建筑师，是普利兹克建筑奖得主，被称为现代主义建筑大师。1974年，他设计了位于华盛顿特区国家广场的美国国家艺术馆东楼。他的另一杰作是中国（香港）银行大厦（简称中银大厦，1989年），这是香港中环地区最为醒目的摩天大楼之一。这座建筑高达三百零五米（1,000.7英尺），其天线塔就高达三百六十七点四米（1,205.4英尺）。从1989年至1992年这是香港乃至亚洲最高的建筑物，也是美国国土以外的第一个打破305米（1,000英尺）的建筑物。

菲利普·科特柳·约翰逊（1906-2005）是一位颇有影响的美国建筑大师。约翰逊给人的印象总是带着厚厚的圆框眼镜，他是几十年来美国最知名的建筑界人物。约翰逊作为实践派建筑师早期的影响是他对玻璃这种建筑材料的广泛使用，他的代表作是玻璃大厦（1949年），这座大厦位于康涅狄格州的新迦南，是为他的个人居所设计的，其影响力也特别广泛。1930年，在他的率领下成立了纽约现代艺术馆的建筑与设计部，后来（1978年）作为受托人，他被授予美国建筑师金奖，并于1979年被授予第一枚普利兹克建筑奖。

理查德·迈耶（1934-）是一位美国建筑设计师，他以其理性化的设计风格和对白色色彩的使用而闻名，这种建筑风格在历史上曾经被用于许多标志性建筑，包括西班牙，意大利南部和希腊这些国家，地中海地区的教堂和粉刷成白色的村庄。1972年，他被人们封为纽约五大建筑师之一。同年，由于参与设计加利福尼亚洲洛杉矶盖蒂中心，他的名声大增，并且建筑风格也一跃成为当时的主流。1984年，迈耶被授予普利兹克建筑奖。2008年，他赢得了艺术与文学学院的建筑金奖。

第五课 亚洲建筑艺术风格

Listening and Speaking 我们承接新项目

A: 我们下一个项目是建一个莎士比亚戏剧影视城。他们想要中世纪的道具。你有什么建议吗？

B: 好吧，我们看看怎么能模仿当时的建筑。我知道当时的建筑使用大量的装饰拱门。

A: 剧中有一幕发生在大教堂，他们希望我们在动工之前提交一些插图。

B: 我也这么想。我认为我们应该设计一下两旁带有扶壁的过道。那样的话人们在教堂里活动时互不干扰。

A: 听起来不错。那么我将开始制作插图，你来搜集资料吧。我们下周同一时间再见面讨论吧。

B: 好的，下周见。

Text A 布达拉宫和紫禁城

公元641年，在迎娶了文成公主后，松赞干布决定建一座豪华的宫殿来作为她的寝宫，并且让他的子孙都铭记这件事。这个宏大的宫殿依红山而建，是历世达赖喇嘛的冬宫，也是过去西藏地方统治者政教合一的统治中心。布达拉宫高117多米（384英尺），宽360米（1180英尺），占地9万平方米。布达拉宫由白宫和红宫组成。前者是非宗教用途，后者为宗教用途。

布达拉宫依山垒砌，群楼重叠，殿宇嵯峨。坚实敦厚的花岗石墙体，松茸平展的白玛草墙领，金碧辉煌的金顶，具有强烈装饰效果的巨大鎏金宝瓶、幢和经幡，交相辉映，红、白、黄三种色彩的鲜明对比，分部合筑、层层套接的建筑型体，都体现了藏族古建筑迷人的特色。

宫殿的设计和建造根据高原地区阳光照射的规律，墙基宽而坚固；墙基下面有四通八达的地道和通风口。屋内有柱、梁、椽木等，组成撑架。铺地和盖屋顶用的是叫"阿尔嘎"的硬土，各大厅和寝室的顶部都有天窗，便于采光，调解空气。宫内的柱梁上有各种雕刻，墙壁上的彩色壁画面积有2,500多平方米。

紫禁城位于北京城的中心，是明清两朝的皇宫。紫禁城呈长方形，占地74公顷，是世界上最大的建筑群。9,999座建筑周围是六米深的护城河和十米高的城墙。

紫禁城开始筹建于1407年（明朝的第三个皇帝明成祖永乐5年），完工于1420年，历时14年。据说，上百万的建造者，其中包括十几万名工匠进行了长期的辛勤劳作。所用的石头是从北京的郊区房山采来的。据说，路上每隔50米就挖一口井，这样冬天就可以把水泼在路上，使之结冰，以便拖动巨大的石头到城里。大量的木材和其他材料是从很远的省份运来的。古代的中国人民在建造紫禁城时展示了他们精湛的技艺。以其红色城墙为例，地基8.6米宽，而顶部是6.66米宽，因此，试图爬上这种尖形的城墙是不太容易的。砖是由白石灰和糯米制成的，水泥是由糯米和鸡蛋清制成的。这种不可思议的材料使城墙格外坚固。

因为黄色是皇族的象征，所以黄色成为了紫禁城的主要颜色。屋顶均有黄色的琉璃瓦，装饰也被漆成黄色的，甚至是地上的砖也是经过特殊工艺制成的黄色。

Text B 日本古京都的建筑

古京都遗址位于前首都平安京区域，始建于公元794年，从那时起到江户时代（1600–1868）一直是日本的首都。古京都仿效古代中国首都形式建造。古京都的最初设计是模仿中国隋唐时代的长安和洛阳，整个建筑群呈长方形排列，以贯通南北的朱雀路为轴，分为东西二京。东京仿照洛阳，西京模仿长安城，中间为皇宫。宫城之外为皇城，皇城之外为都城。城内街道呈棋盘形，东西、南北纵横有秩，布局整齐划一，明确划分皇宫、官府、居民区和商业区。

京都御所是日本的旧皇宫，又称故宫。从奈良迁都到明治维新的1074年中，它一直是历代天皇的住所，后又成为天皇的行宫。京都皇宫位于京都上京区，前后被焚7次，现在的皇宫为孝明天皇重建，面积11万平方米,四周是围墙。

平安神社建筑宏伟壮丽,是明治时代园林建筑的代表作。神社内湖上的亭阁,都是仿照中国西安寺庙的结构建造的。

二条城,建于1603年。它的富丽堂皇与朴素的故宫恰成鲜明对比。城堡以巨石做城垣,周围有东西长500米,南北长300米的护城河,河上有仿唐建筑。

被称为"三步一寺庙、七步一神社"的京都有寺庙1500多座,神社2000多座,这里是日本文化艺术的摇篮,佛教的中心。

第六课 中国古代建筑经典

Listening and Speaking 一次愉快的旅行

A:约翰,听说你去叙利亚进行考古挖掘了。那次经历很难忘吧。

B:是的,这是一项巨大的工程。我们同时有200人做这个工程。

A:是什么激发你们承担了如此浩大的工程?

B:我们的调查是根据去年发现的一些建筑物进行的,以前看到的那些建筑物的建筑风格非常自然,相当简朴。不过当时我们找到的则是很多饰品,珠宝和其他一些带有异族特色的物品。

A:你们发现了很多吗?

B:是的,我们研究发现当时的文明程度似乎很高。该项目成果令人可喜,尤其是我们的主管,他兴奋不已,他真想组织一次盛大的庆典来庆祝一下才罢休。

A:那场面肯定会相当壮观。祝贺你们旅行成功并平安归来!

Text A 灌溉工程和桥梁

纵观中国历史,保存和遗留下来许多值得全世界人民敬佩的、伟大的建筑工程和奇迹。

首先,大运河是古代中国的巨大灌溉工程。它具有1400多年的历史,全长1794公里,是世界最古老的运河之一,也是世界最长的人工河。它的总长远远超出了其他两大运河:苏伊士和巴拿马运河。

大运河北始于北京的通县,南至浙江杭州,沿途连接了五大河:海河、黄河、淮河、钱塘江和长江。在春秋时代晚期,先是在江苏省的扬州开挖,用以将长江水北引。后来在隋炀帝时期通过六年的横征暴敛的开掘才进一步扩展成今天大运河的规模。大运河是元、明、清时期连接中国南北的运输主干道,对于南北经济文化交流起到了巨大的作用——而这个作用是流向自西向东的自然河流所无法比拟的。

其次就是都江堰。它是历史上代表科学技术发展的奇观,也是世界上仍在使用的最古老的水利工程。三国时期蜀国的地方官员李冰于2200年前设计和修建了都江堰,以控制洪水泛滥。除了它的灌溉和控制洪水的功能之外,都江堰也是一个美丽的旅游景点,被联合国教科文组织列为世界文化遗产。它主要包括三大部分:鱼嘴、飞沙和宝瓶口。都江堰将汹涌的河水分成内外两个渠道,外渠作为排水系统用来防洪,内渠将水引入农田进行灌溉。迄今为止,它对我国的农业仍起着至关重要的作用。

最后要谈到的是桥梁。中国是许多桥梁种类的发源地。马可波罗就在他的书中谈到了古城杭州附近的由木头、石头和铁建造的桥梁12000多座。

最著名的铁索桥是四川甘孜的泸定桥。它建于1705年,是横贯大渡河的铁索桥,长100米,宽28米,离水面高度14米。整个桥梁有十三根铁索,每个重2.5吨。其中九根平行的铁索连接两岸,另外四根铁索分悬左右两边起到扶手的作用。

以上提到的这些工程体现了古代中国人民的智慧和能力。在为这些文化遗迹感到骄傲的同时,我们应该努力将我国的杰出民族文化发扬光大。

Text B 颐和园

北京颐和园，建于1750年，在1860年的战争中毁于一旦，并于1886年在原址重建，是中国园林设计的代表作。山和水这些自然景观与亭子、大厅、宫殿、庙宇和桥梁这样的人工景色完美结合，形成了具有非凡美学价值的和谐整体。

颐和园，以长寿宫和昆明湖为主体，占地面积2.9平方公里，其中四分之三被水覆盖。各种宫殿、花园及其他的古建筑占据了70000平方米的建筑面积。因藏有大量的价值连城的文化古迹，颐和园名列中国国家特殊保护历史文化遗迹之首。

无论从园林的设计还是建造上来说，颐和园都是中国古典建筑的丰碑。它通过借景于周边景色，以天衣无缝的结合不仅体现了皇家园林的壮丽，也体现了自然之美。这种结合完美地体现了传统中国园林设计的指导思想：人工与天然相结合。1998年12月，联合国教科文组织以以下评论将颐和园列为世界文化遗产：1.北京颐和园是中国园林设计艺术的杰出表现，将自然与人工和谐统一；2.颐和园是中国园林设计思想与实践的缩影，在东方园林设计发展中扮演了重要角色。3.以颐和园为代表的皇家园林，是世界文明的重要代表。

第七课 古代建筑艺术简史

Listening and Speaking 旅途致谢

A：你好，李先生。我是专程前来对你这些天的帮助表示感谢的。我在这里过得很愉快。
B：不用客气。
A：我今天就要回美国了。非常感谢您在我停留期间给予的帮助。
B：别客气。我十分高兴能帮上忙。
A：如果将来有什么需要我帮忙的话，找我好了。
B：谢谢。祝您旅途顺利。
A：我会的。您保重。

Text A 埃及金字塔

有史以来，古代埃及金字塔就吸引着无数旅游者和探险家前往参拜。时至今日仍有很多旅游者、数学家和考古学家不断前往探险并进行考古研究。

最大、最著名的金字塔是位于吉萨的大金字塔，由斯奈夫鲁的儿子胡夫，（也被称为基奥普斯后者是他的希腊名）建造。金字塔占地面积13英亩，三角面斜度51度52分（约为52度）、底座边长755英尺。原高481英尺，现高450英尺。科学家估计，它的每块石块平均有2吨重，最大的重达十五吨。其他两个主要的金字塔也建在吉萨，是分别为胡夫的儿子卡夫拉国王和卡夫拉的继任者孟考拉建造的。孟考拉金字塔也位于吉萨，它就是著名的狮身人面像。这座大型雕像由一个狮身雕像和一个人的头部雕像结合而成，雕刻于卡夫拉时代。

关于金字塔的建造一直有很多的猜测。古埃及人可能用凿子、钻、锯等铜器来切割相对较软的石材。更加难以处理的硬花岗岩则用于建造墓室的墙壁及部分外墙。工匠们或许用沙子等研磨粉配合钻、锯来切割石料。确定金字塔的主中心点需要掌握天文学知识，而且，当时可能采用了注水沟槽用于控制金字塔周边的水平度。有一幅壁画，上面画有工人们利用放在用水冲滑的地面上的滑板来移动巨型雕像的情景，这向人们展示了当时工人们是如何搬运巨石的。在金字塔内部通过坡道将石块运到它所在的位置。最后,外部塔身的石块沿着斜坡由上而下放置,完工后坡道被拆除。

金字塔的建造从第四到第六王朝时达到鼎盛。一千多年来,埃及人不断地在建造一些小型的金字塔。其中有几十个已经被发现了,但其他的可能仍然埋在沙子中。显而易见的是金字塔没有发挥对国王木乃伊的保护作用,反倒成了盗墓者显见的目标,于是,后来国王就被埋葬在岩壁中隐藏的墓穴里。尽管宏伟的金字塔没有把建造它们的埃及国王的遗体保护起来,但金字塔仍然使国王们的名字以及他们的故事得以流传到今天。

Text B 世界著名古建筑遗迹

(1) 吴哥窟

吴哥窟是九至十三世纪12个高棉(谷美尔)国王所建造的,吴哥在东南亚占地超过120平方英里,包括几十个主要的建筑遗址。公元802年开始修建吴哥窟的时候,苏耶跋摩二世凭借做大米和货物贸易积累财富,夺取了王位,开创了空前的艺术与建筑成就时期。吴哥窟遗址是古代帝国的中心,是这一时期建筑物的典型代表。在令世人称奇的金字塔和曼陀罗形的神庙中,位于柬埔寨丛林里的吴哥窟是迄今为止世界上现存的最大的寺庙。它结构复杂,神社内人物肖像绘画惟妙惟肖,极具宗教色彩。今天,许多国家都想尽自己的力量保护和修缮这些寺庙,但是这项工作始终进展缓慢。

(2) 玛雅城

寺宙金字塔是古代玛雅城最显著的特征。它们是用手工切割的石灰石构筑的,并耸立在周围的所有建筑群中。虽然寺庙本身通常包含一个或多个房间,但房间是如此狭窄,因此它们只能在祭祀场合使用,而不能供公众市民使用。在整个建筑群中,礼仪中心的整体规划最为重要。

玛雅建筑的典型特征包括用托臂支撑的拱顶石和穹顶。玛雅金字塔用托臂支撑的拱顶石很像欧洲的拱门建筑,没有塔基,塔的最顶端是一道狭窄的三角形的缺口,不像拱门那么宽敞。有人认为这种造型说明玛雅人当时还没掌握打地基的技术;但是,也有人认为玛雅人是有意为之:托臂拱顶石共分九层,代表着苍穹大地的九级阶梯,再加上塔顶石基平台共十层。这体现了玛雅人当时就会运用天文学。

或许玛雅的建筑师觉得寺庙不够大,因此增加了上部的延伸。条脊高高的,装饰着绘有灰泥的浮雕,如同寺庙的外墙。有同样装饰的还有门口,大门柱和许多其他玛雅建筑的外墙,同时还用石刻或木刻装饰。

(3) 泰姬陵

几乎每个人都阅读过有关印度的泰姬陵的故事。它是世界上最美丽的建筑物之一。三百多年前,沙贾汗为他的妻子修建了坟墓——泰姬陵。

沙贾汗想把妻子的坟墓造得完美无缺,他不在乎时间和金钱。因此,他召集了全亚洲2万多名工匠,并用了17年的时间才完成这项浩大的工程。

该建筑坐落于红色的砂岩平台上,四座白色尖塔从平台的角落拔地而起,建筑物的中心有一个大圆顶,围绕这个大圆顶有四个较小的圆顶。该建筑是由精细的白色和彩色大理石建成的,它有八个边和许多拱门。

泰姬陵坐落在一个美丽的花园之中。绿色的树木使大理石显得更加洁白。在主入口前面还有一个又长又窄的水池。

第八课 生态与建筑

Listening and Speaking 向志愿者求助

A: 下午好,女士。有什么需要我帮忙的吗?
B: 哦,是的,谢谢。可以帮我把这些行李搬到那边大门口吗?
A: 没问题。
B: 你人真好。
A: 我很高兴能帮上忙。
B: 我还想问问怎样去机场?
A: 您可以坐公共汽车去。汽车站就在转角处。
B: 我到那儿要多久?
A: 我想大约要一个小时。
B: 我要赶4点的航班,恐怕来不及。
A: 这样的话,我建议您打出租车去。
B: 好主意。
A: 如果您不介意的话,您可以在这里看着行李,我帮您叫出租车。
B: 那真是太感谢您了。
A: 不客气。

Text A 城市让生活更美好

展示、论坛与活动是2010年上海世博会的三大组成部分,三者都围绕"城市让生活更美好"世博会这一核心主题展开。现今,半数的世界人口居住在城市中,预计在未来的20年中,60%的世界人口将成为城市居民。在城市化的进程中,人类面临诸多问题与挑战。

信息化与城市发展

21世纪是全球城市化的世纪,城市在国家或地区中的作用将越发重要。当前,城市的传统要素正随着信息化的演进进行着快速的重新塑造与组合,其社会经济基础也正发生重大转变。随着信息通信技术(ICT)的迅猛发展,人类已迈入信息社会。信息化通过利用高效、集成的现代通信和信息网络技术,对资金、人才、设备等物质要素进行重新整合,强化各种生产要素向城市迅速聚集,带动城市空间结构与空间功能的优化配置,推进着城市向前发展。城市信息化是未来城市的共同选择,也是城市合作的热点。

科技创新与城市未来

自人类创造城市以来,城市的发展与科技的创新始终相互促进、互为依托,历届世博会更是各国最新科技成果的集中展示。2010年上海世博会将秉承世博会的传统,贯穿人、城市与地球和谐共处的理念,汇聚全球科技,集中人类智慧,共同演绎"城市,让生活更美好"的本届世博会主题。

环境变化与城市责任

"全球思考,城市行动"是长久以来环保行动的口号之一。城市既是各类环境问题的始作俑者,也是环境恶化和气候变化的主要承担者和"利益攸关方"。环境保护、气候变化和节能减排等重大问题的解决,需要政府、企业和公民这三大城市责任主体的共同努力。

和谐城市与宜居生活

为了更好的生活,人类定居城市。然而在全球各地,在享受城市发展馈赠的同时,城市的居民也承受不同阶段城市化进程所带来的诸多问题和困扰。城市时代的到来,使得城市环境及其可持续保障愈来愈成为人类宜居生活的重要载体。如何通过更为有效的手段使得城市更好地满足人类生活的多元要求,从而真正实现"城市让生活更美好"的目标,如何制定更为长效的机制来保障城市和谐发展和市民的宜居,从而实现全球人类社会的福祉,这些问题都将在本届世博会上寻求到可能解决的方案。

Text B 科纳卡光电塑料

2001年科纳卡科技公司创建之初的憧憬:想象一个无二氧化碳排放的世界;即使在最偏远最贫困地区的村庄也有互联网接入,每一户人家都能亮灯;一个能源丰富、安全并且是真正的绿色环保的世界。每一天,他们都使憧憬与现实更加接近。该公司的科纳卡光电塑料重量轻,柔韧性、伸缩性、扩展性和适应性都很好,适用领域广泛,在商业、工业、政府和普通消费者中广泛应用。

科纳卡光电塑料是一种光电材料,它能捕捉室内和室外光并且转换成直流电(DC)的电能。这种能量可以立即投入使用,或存储起来供日后使用,也可转换为其他形式。这种塑料可以应用到许多有发展潜力的设备中去:从微电子产品到便携式电源,远程电源以及建筑集成的应用。

科纳卡公司的光电塑料与其他有机光电技术相比,拥有以下几大优势:

电力塑料的可调和的细胞化学物质能吸收特定波长的光——就像光波谱一样;与传统的有机光电技术不同,科纳卡公司光电塑料,从日出到日落,可以收集离太阳光直射轴最高可达70°以内的光能,并随时可收集到室内的光能。

科纳卡光电塑料是唯一利用一切可循环再造材料的塑料薄膜光电技术。电力塑料应用可分为四大类:微电子,便携式电源,远程电源和建筑(光伏)集成的应用。目前,我们的制造合作伙伴正在开发的应用产品包括:

便携式电池充电器——给笔记本电脑,手机和灯笼充电。

微电子——如传感器,智能卡,遥控启动器

个人护理设备——如电动剪(电动剃须刀)和电动牙刷

户外装备——如帐篷,背包——可以驱动便携式电子设备和照明

应急发电机——这将使警察,军队和紧急救援人员维持重要通信

遮盖物结构——在任何有阳光的地方可以为个人电子设备充电

车库顶篷——可以匀速缓慢地给电子汽车充电

窗帘和综合窗口面板——将捕获太阳能,并将其转换为办公室及住宅用电

在绿色环保和可再生能源领域,太阳能的地位无可替代。科纳卡公司的突破性技术正在将明天的奢望转化为今天的实惠制造材料。

随着光电塑料的应用,个人将会拥有独立的能源。想象一下,在你走路去上班的路上,背包可以给装在里面的笔记本电脑充电,公文包可以给你的手机充电,手袋能给你的相机补给电量;想象一下,办公室窗户能给你工作地点的电脑和咖啡壶充电;想象一下,当你在海滩上享受美好时光的时候,在远离高压电网的地方,一把遮阳伞就能够让你保持旺盛的精力。想象一下,您在开车的时候汽车的顶篷能给您的电子设备充电,汽车的外罩能给您的电动汽车充电,一顶帐篷可以在日落时打开您的阅读灯并温暖您的睡袋。

Appendix 3 Glossary 总词汇表

A

abate [ə'beit] v. make less active or intense; become less in amount or intensity 减少, 减弱, 减轻

abrasive [ə'breisiv] n. a substance that abrades or wears down 研磨剂 adj. causing abrasion 磨平的

absorb [əb'sɔ:b] v. take (sth) in; suck up; include (sth/sb) as part of itself or oneself, incorporate; merge with 吸收, 理解, 掌握, 使全神贯注

accelerate [ək'seləreit] v. cause to move faster 加速, 促进, 增加 (1)The leader is losing ground as the rest of the runners accelerate. 领先者在其余赛跑者加速时就逐渐失去了优势。(2)The car accelerated as it overtook me. 那辆汽车一加速就超越了我。(3)The driver stepped on the gas and accelerated the car. 司机加大油门, 让汽车加速行驶。

access ['ækses] n. a way of entering or reaching a place; the opportunity or right to use sth or to see sb/sth 通道, 通路, 机会, 权利 (1)The only access to the farmhouse is across the fields. 去那农舍的唯一通路是穿过田野。(2)The police gained access through a broken window. 警察从一扇破窗户里钻了进去。(3)You need a password to get access to the computer system. 使用这个计算机系统需要口令。

accommodate [ə'kɔmədeit] v. to supply or provide, esp. with lodging or board and lodging (1)为(某人)提供住宿(或膳宿、座位等) (2)容纳, 为……提供空间 (3)考虑到, 顾及 (4)accommodate sb(with sth) 帮忙, 给……提供方便 (5)accommodate to sth; accommodate sth/yourself to sth 顺应, 适应(新情况) I quickly needed to accommodate to the new schedule. 我需要迅速适应新的时间表。

account for 说明, 占, 解决, 得分 Cotton accounts for 70% of our export. 棉花占我们出口的70%。

according to 遵守, 依照 by right; according to law 依据正义, 按照法律

acre ['eikə] n. a unit of area (4,840 square yards) used in English-speaking countries 一英亩约为4050平方米 (1) 3,000 acres of parkland 3000英亩开阔绿地 (2) a three-acre wood 一片三英亩的林地 (3) Each house has acres of space around it. 每座房屋四周都有大量空地。

AD ['ei 'di:] n. used in the Christian calendar to show a particular number of years since the year when Christ was believed to have been born (from Latin Anno Domini) 公元(源自拉丁语 Anno Domini)

admire [əd'maiə] v. to regard with pleasure, wonder, and approval; to have a high opinion of; esteem or respect 赞美, 钦佩, 羡慕 (1) I really admire your enthusiasm. 我确实钦佩你的热情。(2) You have to admire the way he handled the situation. 你不得不佩服他处理这个局面的手段。admiring [əd'maiəriŋ] adj. 赞赏的

adopt [ə'dɔpt] v. ~ sb (as sth) take sb into one's family, esp. as one's child or heir 收养某人(尤指作为儿女或继承人), 过继 ~ sb as sth choose sb as a candidate or representative 挑选某人作候选人或代表

aesthetic [i:s'θetik] adj. relating to the philosophy or theories of aesthetics; of or concerning the appreciation of beauty or good taste characterized by a heightened sensitivity to beauty; artistic 美学的, 美的, 有审美感的

afford [ə'fɔ:d] v. be able to spare or give up [常与 can, could, be able to 连用] 担负得起费用(损失、后果等), 花费得起, 经受得住, 抽得出(时间), 给予, 提供 (1)They walked because they couldn't afford (to take) a taxi. 他们因为坐不起计程车而步行。(2)You can't afford £90. 你可不能花90英镑。(3)I'd love to go on holiday but I can't afford the time. 我倒想去度假, 可是抽不出时间来。(4)We would give more examples if we could afford the space. 假如我们能匀出篇幅来, 就可以多举些例子了。

affordable [ə'fɔ:dəbl] adj. low-cost, low-priced, heap, inexpensive 负担得起的, 普及型的

agriculture ['ægrikʌltʃə] *n*. the science, art, and business of cultivating the soil, producing crops, and farming 农业

aim to 目标在于……，打算，以……为目标
aim to reduce road traffic accidents 目的为减少道路交通事故。

aisle [ail] *n*. a passageway separating seating areas in a theatre, church, etc; gangway（教堂的）走廊，侧廊，耳堂 (1)He followed the usher down the aisle. 他跟着领座员沿着通道走过去。(2)The organ play as the bride come down the aisle. 当新娘沿着通道走过来时，风琴演奏了起来。(3)The girl ushered me along the aisle to my seat. 引座小姐带领我沿着通道到我的座位上去。(4)Things reached the point where the two groups sat on opposite sides of the church, glaring across the aisle. 事情发展到了两群人马各坐在教堂一侧，隔着通道怒目相视的地步。

alignment [ə'lainmənt] *n*. the act of adjusting the parts of a device in relation to each other 调整成直线

along with 连同……一起，与……一道，随同……一起 Can you come along with us tomorrow? 你明天能来和我们一起去吗？

altar ['ɔːltə] *n*. (in Christian churches) table on which bread and wine are consecrated in the communion service 圣餐桌; table or raised flat-topped platform on which offerings are made to a god 供桌, 祭坛

altitude ['æltitjuːd] *n*. the height above sea level 海拔，海拔高度 a place that is high above sea level（海拔高的）高处，高地 (1)The plane was then flying at an altitude of 8,000 feet. 当时飞机在八千英尺的高度飞行。(2)What is the altitude of this village? 这个村子海拔多少？ (3) We are flying at an altitude of 20,000 feet. 我们的飞行高度是两万英尺。(4)He found it difficult to breathe at high altitudes. 他觉得在高空呼吸困难。

angular ['æŋgjulə] *adj*. lean or bony; having an angle or angles 1. 瘦骨嶙峋的,骨瘦如柴的 (1) an angular face 瘦削的 (2)a tall angular woman 又高又瘦的女人 2.有棱角的，有尖角 an abstract design of large angular shapes 大棱角形的抽象图案 3.笨拙的，生硬的 He walked with angular strides. 他迈着僵硬的步伐行进。

antiquarian [ˌænti'kweəriən] *n./adj.* concerned with the study of antiquities or antiques 古物研究者，收藏古物的

antisepsis [ˌænti'sepsis] *n*. (of non-living objects) the state of being free of pathogenic organisms 防腐，消毒

analogous to 类似的，可比拟的 The heart is analogous to a pump. 心脏和水泵有相似之处。

aperture ['æpətʃə] *n*. (fml 文) a small hole or space in something, narrow opening 窄孔，隙缝

apiece [ə'piːs] *adv*. to or from every one of two or more (considered individually) 每人，每个，各 (1)Yorke and Cole scored a goal apiece. 约克和科尔各进一球。(2)The largest stones weigh over five tones apiece. 最大的石头每块重五吨以上。

applique [æ'pliːkei] *n*. a decorative design made of one material sewn over another 嵌花，贴花，缝花 *adj*. 嵌花的

apply... to 应用于

approximately [ə'prɔksimitli] *adv*. 大约，大致，近乎

approximate [ə'prɔksimit] *adj*. an approximate number, amount, or time is close to the exact number, amount etc, but could be a little bit more or less than it [= rough; ≠ exact]: 近似的，大约的 What is the approximate number of students in each class? 每个班的学生大约多少人？ These percentages are only approximate. 这些只是近似值。 *v*. to be close to a particular number; to be similar to but not exactly the same as something: 近似，接近 This figure approximates to a quarter of the UK's annual consumption. 这个数字接近英国年消费量的四分之一。

Aquarius [ə'kwɛəriəs] *n*. a zodiacal constellation in the S hemisphere lying between Pisces and Capricorn on the ecliptic 宝瓶（星）座，宝瓶宫（黄道第十一宫），属宝瓶座的人（约出生于1月21日至2月19日）

archaeological [ˌɑːkiə'lɔdʒikəl] *adj*. related to or dealing with or devoted to archaeology 考古学的，考古上的
(1)From archaeological evidence it is apparent that each group superimposed its ideas and culture on the

previous one. 从考古事实看，每一群体都将自己的观点和文化溶合于前代群体中。(2)A town of ancient Palestine north of Jerusalem, it is now a major archaeological site. 贝瑟尔巴勒斯坦—古老城镇，位于耶路撒冷以北，是现在的一个主要考古地区。

archeologist [ˌɑ:kiˈɔlədʒist] *n.* an anthropologist who studies prehistoric people and their culture 考古学家

archer [ˈɑ:tʃə] *n.* person who shoots with a bow and arrows, esp. as a sport or (formerly) in battle 射箭运动员，（旧时）弓箭手，【天】（大写）人马座[the S]

architect [ˈɑ:kitekt] *n.* someone who creates plans to be used in making something (such as buildings) 建筑师，设计师

architectural [ɑ:kiˈtektʃərəl] *adj.* of or pertaining to the art and science of architecture 建筑学的，建筑上的

archway [ˈɑ:tʃwei] *n.* a passageway under a curved masonry construction 拱门，拱道

architecture [ˈɑ:kitektʃə] *n.* the art and science of designing and superintending the erection of buildings and similar structures; a style of building or structure 1.建筑学 to study architecture 学习建筑学 2. 建筑设计，建筑风格 (1)the architecture of the eighteenth century 十八世纪的建筑风格 (2)modern architecture 现代建筑设计 (3)He likes Greek architecture. 他喜欢希腊建筑式样 3.体系结构，（总体、层次）结构 4.建筑物 This is the most impressive architecture I've seen on this trip.这是我此次旅行见到的最令人难忘的建筑。

artery [ˈɑ:təri] *n.* any of a branching system of muscular, elastic tubes that carry blood away from the heart to the cells, tissues, and organs of the body; a major route of transportation into which local routes flow 动脉，要道 1. 动脉 blocked arteries 被阻滞的动脉 2.干线（主要指公路、河流、铁路线等）This is the place where the three main arteries of West London traffic meet. 这是伦敦西区三条主要干线会合的地方。

artificial [ˌɑ:tiˈfiʃəl] *adj.* made by human beings; produced rather than natural 1.人工的，人造的，假的(1)an artificial limb / flower / sweetener / fertilizer 假肢，假花，人造甜味剂，化肥(2)artificial lighting / light 人工照明，人造光 2.人为的，非自然的 (1)A job interview is a very artificial situation. 求职面试是一个相当不自然的场面。(2)the artificial barriers of race, class and gender 种族／阶级／性别 的人为障碍 3.虚假的，假装的(1)artificial emotion 假装的情感(2)Her artificial manner made me sick. 她那装腔作势的样子令我恶心。

astronomy [əˈstrɔnəmi] *n.* the branch of physics that studies celestial bodies and the universe as a whole 天文学

as if 好像，仿佛

as such 同样地 He is a child,and must be treated as such. 他是个孩子，必须被当作孩子对待。

as well 倒不如，还是……的好，最好……还是

as well as 也，又 Action as well as thought is necessary. 行动与思考同属必要。

a succession of 一连串的 There are a succession of rainy days here. 这里一连好几天都是雨天。

attractive [əˈtræktiv] *adj.* appealing to the senses or mind through beauty, form, character, etc; arousing interest 有吸引力的(1)an attractive woman 妩媚的女人 (2)I like John but I don't find him attractive physically.我喜欢约翰，不过我认为他长得并不英俊。

austere [ɔ:ˈstiə] *adj.* severely simple or plain,serious 严峻的，苛刻的(1)Father was an austere man, very strict with his children.父亲是个严厉的人,对子女十分严格。(2)They lived an austere life. 他们过着极端清苦的生活。(3)I like an austere style of writing. 我喜欢朴实无华的文体。

automobile [ˈɔ:təməubi:l, ˌɔ:təməˈbi:l] *n.* a motor vehicle with four wheels; usually propelled by an internal combustion engine 汽车

a variety of 各种各样的……, 种种……There is a variety of flowers in the garden.

available [əˈveiləbl] *adj.* obtainable or accessible and ready for use or service; not busy; not otherwise committed

可得到的,可用的,可与之联系的,可获得的,可购得的,可找到的 available resources / facilities 可利用的资源 / 设备, readily / freely / publicly / generally available 随时 / 随手 / 公开场合 / 一般可以得到的,有空的,有效的 Will she be available this afternoon？今天下午她有空吗？

aversion [ə'və:ʃən] *n.* a feeling of intense dislike (1)aversion to (sb / sth) 厌恶,憎恶 (2)a strong aversion 深深厌恶(3)feelings of aversion and disgust 憎恶之情

axis ['æksis] *n.* (pl. axes['æksi:z]) 1.轴(旋转物体假象的中心线)2.(尤指图表中的)固定参考轴线,坐标轴 the vertical / horizontal axis 纵 / 横坐标轴 3. (透镜的)光轴 4.(几何)对称中心线(将物体平分为二)(1)an axis of symmetry 对称轴 (2)The axis of a circle is its diameter. 圆的对称中心线就是直径。5.轴心(国与国之间的协议或联盟)the Franco-German axis 法德轴心

American Institute of Architects 美国建筑师学会 该学会是美国建筑界权威性的组织,总部位于华盛顿哥伦比亚特区。

B

be based on 根据,建立在……基础上 based on experience; empirical. 根据经验的,全凭观察和实验的

be bent on 一心想要,决心要 He is bent on mastering Spanish. 他一心一意想要精通西班牙语。

be composed of 由……组成 The Election Committee shall be composed of 800 members. 选举委员会将由800人组成。

be divided into 把……分隔成若干部分 The investigator will be divided into three groups. 所有的调查员将被分为三个小组。

be used to 被用来 Lasers can be used to perform operations nowadays. 现在激光可以用来做手术。

be worth doing 值得…… If something is worth doing, it is worth doing well. 如果事情值得做,就值得做好。

be combined with 与……相结合 In sum, theory must be combined with practice. 总之,理论要与实践相结合。

be confronted with 面临,面对,对照 The new system will be confronted with great difficulties at the start 这种新的制度.一开始将会面临很大的困难。

be necessary to 为……所必需 It may be necessary to buy a new one. 也许有必要买个新的了。

be well known for 因……而著名 He is well known for fair dealing. 他以平等待人著称。

bring up 提高,达到(to)The road will bring you up to the top of the cliff. 这条道路通向悬崖的顶端。

be renowned for 以……闻名 Suzhou is renowned to the world for its arts and crafts. 苏州的工艺品闻名全球。

be drawn to 被……所吸引 You will be drawn to the buildings you see while walking along the streets in Shanghai. 走在上海的街道上,你会被两边的建筑物所吸引。

be crowded out 被排挤 Small shop is crowded out by the big supermarket. 小商店受到大型超市的排挤。

be regarded as... 被认为是…… The old man does not want to be regarded as a burden. 这位老人不想被当作负担。

BC ['bi: 'si:] *n.*before Christ (used in the Christian calendar to show a particular number of years before the year when Christ is believed to have been born)公元前(基督教会历法用)

beam [bi:m] *n.* a long thick straight-sided piece of wood, metal, concrete,etc, esp one used as a horizontal structural member 1.光线,(电波的)波束,(粒子的)束 2.梁 The cottage had an original fireplace and exposed oak beams. 小屋有一个昔日的壁炉和裸露的橡木梁。3.平衡木 The gymnast performed a somersault on the beam. 体操运动员在平衡木做了个空翻。4. off beam 不正确,错误

board [bɔ:d] *n.* a long wide flat relatively thin piece of sawn timber; a group of people who officially administer a company, trust, etc 1. 板,(尤指)木板 2.(尤用于构成复合词)用木板(或板材)3.董事会,委员会,理事

会 4.伙食,膳食费用 5. across the board 全体 6.be above board（尤指商业安排）开诚布公 7.go by the board（计划或原则）被废弃,被忽视 8.on board 在船上（或飞机上、火车上）9.take sth on board 采纳,接纳(主意、建议)

Bourges 博格斯,位于奥尔良东南偏南地区法国中部城市,曾是奥古斯都时期的罗马首府。

bracket ['brækit] *n.* a support projecting from the side of a wall or other structure 支柱,托架 1.括号 (1) Publication dates are given in brackets after each title. 出版日期括于书名后面。(2)Add the numbers in brackets first. 先把括号里的数字加起来。2.(价格、年龄、收入等的)等级 (1)Most of the houses are out of our price bracket. 大多数房子都超出我们的价格范围。(2)people in the lower income bracket 低收入等级的人们 3.(固定在墙上的)托架,支架

brilliant ['briljənt] *adj.* a. shining with light; sparkling; outstanding;exceptional 1.巧妙的,使人印象深的(1) What a brilliant idea！真是个绝妙的主意！(2)a brilliant performance / invention 出色的表演,杰出的发明 2.很成功的 a brilliant career 一帆风顺的事业 3.聪颖的,技艺高的 4.明亮的,鲜艳的 brilliant sunshine 明媚的阳光 5.很好的

Buddhism 佛教

burial ['beriəl] *n.* the ritual placing of a corpse in a grave 埋葬

buttress ['bʌtris] *n.* also called pier, a construction,usually of brick or stone,built to support a wall【建】撑墙,扶壁,(前)扶垛,支[肋]墩~ sth (up) support or strengthen sth 支持或加强某事物(1)He buttressed up the argument with lots of solid facts. 他以大量确凿的事实支持这个论点。(2)You need more facts to buttress up your argument. 你需要有更多的事实来支持你的论据。(3)More government spending is needed to buttress industry. 为加强工业发展需要政府增加拨款。

by contrast 相比之下

Badaling 八达岭

Beijing Courtyard Houses 北京四合院

Book of Changes《易经》

C

calligraphy [kə'ligrəfi] *n.* handwriting, esp. beautiful handwriting considered as an art 善于书写,书法,笔迹

Cambodia [kæm'bəudiə] *n.* a nation in southeastern Asia 柬埔寨(亚洲)

canal [kə'næl] *n.* an artificial waterway or artificially improved river used for travel, shipping, or irrigation 1.运河,灌溉渠 the Panama / Suez Canal 巴拿马 / 苏伊士运河 an irrigation canal 一条灌溉渠 2.气管,食道(动植物体内的)管道 3.【天】(像水星表面上的)透过天文学望远镜看到的朦胧线条

capture ['kæptʃə] *n.* the act of catching, taking, or winning, as by force or skill.抓取, 战利品, 捕获之物 v.to attract and hold 捕获,记录, 接收, 拍摄, 用武力夺取, 攻取 The company has now captured almost 90% of the market. 这家公司现已占有90%的市场份额。1. to capture sb's attention / imagination / interest 引起（注意、想象、兴趣）2. to capture sb / sth on film / tape / canvas 拍摄,录制,绘制 3. to capture sb's heart 使……爱上(或倾心)

cardinal ['kɑ:dinəl] *n.* a priest of high rank in the Roman Catholic Church 红衣主教,a variable color averaging a vivid red 鲜红色, a cardinal number 基数 *adj.* serving as an essential component 主要的,最重要的 Respect for life is a cardinal principle of English law. 尊重生命是英国法律最重要的原则。

care about 关心 I don't care about the price, so long as the car is in good condition. 我不计较价钱,只要车的性

能好就行。

carport ['kɑ:pɔ:t] *n.* an open-sided shelter for an automotive vehicle, which consists of a roof supported by posts or walls [美]简陋的汽车棚

carry forward 发扬光大 We should carry the revolutionary tradition forward. 我们应该把革命传统发扬光大。

catapult ['kætəpʌlt] *v.* to shoot forth from or as if from a catapult 猛投 (1)She was catapulted out of the car as it hit the wall. 汽车撞墙时,她被甩出车外。(2)The movie catapulted him to international stardom. 这部电影使他一跃成为国际明星。

cathedral [kə'θi:drəl] *n.* the principal church of a diocese, containing the bishop's official throne 大教堂 (1)This street offers a fine vista of the cathedral. 这条街的尽头是个大教堂,远远望去非常好看。(2)The bishops, priests and deacons processed into the cathedral. 主教、司铎以及助祭列队走进大教堂。(3)Photography is strictly forbidden in the cathedral. 教堂内严禁摄影。(4)The area which a bishop administers has one cathedral and many smaller churches. 一个主教所掌管的地区包括一个大教堂和许多小教堂。(5)I don't like the cathedral style of his writing. 我不喜欢他官腔十足的文风。

cave dwellings 窑洞

cement [si'ment] *n.* a fine grey powder made of a mixture of calcined limestone and clay, used with water and sand to make mortar, or with water, sand, and aggregate to make concrete 1.水泥 2.(干燥后硬化的)水泥 a floor of cement = a cement floor 水泥地板 3.胶合剂,胶接剂,黏固剂 dental cement 压黏固剂 4.(共同利益联合起来的)纽带,聚力 values which are the cement of society 具有社会凝聚力的价值观念

ceremonial [ˌseri'məunjəl] *n.* a formal event performed on a special occasion 仪式,礼仪,礼节 The visit was conducted with all due ceremonial. 访问按照一切应有的礼仪进行。 *adj.* marked by pomp or ceremony or formality 正式的,礼仪/节的

chamber ['tʃeimbə] *n.* a natural or artificial enclosed space 房间 1.会议厅 (1)The members left the council chamber. 议员们离开了会议厅。(2)the Senate / House chamber 参议院/众议院会议厅 2.(议会的)议院 (作特定用途的)房间,室 (1)a burial chamber 墓室 (2)Divers transfer from the water to a decompression chamber. 潜水员从水里转入减压舱。3.(人体、植物或机器内的)腔,室

channel ['tʃænl] *n.* a broad strait, esp. one that connects two seas; the deeper part of a river or harbor, esp. a deep navigable passage 1.电视台 2.频道,波段 3.途径,渠道,系统 (1)Complaints must be made through the proper channels. 投诉必须通过正当途径进行。(2)The newsletter is a useful channel of communication between teachers and students. 简讯是师生之间沟通的有益渠道。(3)The company has worldwide distribution channels. 这家公司拥有遍布全世界的销售网络。4.海峡 the English Channel 英吉利海峡

charger ['tʃɑ:dʒə] *n.* one that charges, such as an instrument that charges or replenishes storage batteries 充电器,装[加]料机,加液器(1)horse used in combat 战马 (2)Please put my charger into your pocket. 请把我的充电器放到你的口袋里。

Cheops ['ki:ɔps] *n.* 基奥普斯(Khufu的希腊名)

chessboard ['tʃesbɔ:d] *n.* a square board divided into 64 squares of two alternating colours, used for playing chess or draughts 棋盘

chisel ['tʃizl] *n.* an edge tool with a flat steel blade with a cutting edge 凿子 *v.* engage in deceitful behavior; practice trickery or fraud 用凿子刻,雕,凿(1)A name was chiseled into the stone. 石头上刻着一个人名。(2)She was chiselling some marble. 她在雕刻一些大理石。

circulation [sə:kju'leiʃən] *n.* passing of something from one person or place to another; spread 流通,循环,传播

(1)We observed the circulation of blood in frogs. 我们观察了青蛙的血液循环。(2)Harvey discovered the circulation of the blood. 哈唯发现了血液循环。(3)This paper has a circulation of more than a million. 这一报纸的发行量达一百多万份。(4)The newspaper's circulation has trebled since last year. 去年以来该报的发行量已增加到原来的三倍。(5)The old coins have been withdrawn from circulation. 旧硬币已经收回不再流通了。(6)Many forged notes are in circulation. 许多假钞在流通。

civilization [ˌsivilaiˈzeiʃ(ə)n] *n.* the type of culture and society developed by a particular nation or region or in a particular epoch;an advanced state of intellectual, cultural, and material development in human society, marked by progress in the arts and sciences, the extensive use of writing, and the appearance of complex political and social institutions 文明，文化，文明社会

civilize [ˈsivilaiz;ˈsivəlaiz] *v.* to educate or improve a person or a society; to make sb's behavior or manners better 教化、开化、使文明，使有教养(1)His wife has had a civilizing influence on him. 他妻子对改进他言谈举止有潜移默化的影响。(2)Many a rough man has been civilized by his wife. 许多粗野的男人在妻子的开导下变得文雅了。

coin [kɔin] *n.* a piece of metal, usually flat and round, that is used as money 硬币 *v.* to fabricate or invent (words, etc)设计，发明，杜撰（新词、习语）(1)The government has decided to coin more 50-penny coins.政府决定铸造更多的五十便士硬币。(2)It was Anderson who coined the word. 是安德生造了这个字。

coincide with 与……相一致 My free time does not coincide with his. 我有空时他没空。

classicize [ˈklæsisaiz] *v.* to make classic, to imitate classical style模仿古典

cliff [klif] *n.* a steep high face of rock（常指海洋边的）悬崖,峭壁 (1)the cliff edge / top 悬崖边缘 / 顶端 (2)the chalk cliffs of Southern England 英格兰南部的白垩质峭壁 (3)a castle perched high on the cliffs above the river 高高耸立在临河峭壁上的城堡

collapse [kəˈlæps] *v.* (break into pieces and) fall down suddenly（破碎并）突然倒塌, 坍塌, 塌陷

colossal [kəˈlɔsəl] *adj.* so great in size or force or extent as to elicit awe. 巨大的，庞大的，广阔的 The singer earns a colossal amount of money. 那歌手现在可赚大钱了。

column [ˈkɔləm] *n.* 圆柱 1.柱，（通常为）圆形石柱，纪念柱(1)The temple is supported by marble columns. 这座庙宇由大理石支撑。(2)Nelson's Column in London 伦敦的纳尔逊纪念碑2.圆柱形物，柱子、柱形物 3.（书、报纸印刷页上的）栏，列(1)He always reads the sports column in the Times. 他总是阅读《泰晤士报》的体育栏。(2)Put a mark in the appropriate column. 在适当的栏里标上记号。(3)Their divorce filled a lot of column inches in the national papers.他们的离婚引起了多家全国报纸的关注。4. (报纸,杂志的)短评栏，专栏,栏目 I always read her column in the local paper. 我一直读她在当地报纸上的专栏文章。

comb [kəum] *n.* a flat device with narrow pointed teeth on one edge; disentangles or arranges hair 1.梳子，篦子（作为装饰物的)发插 3.(公鸡的)鸡冠,鸡冠状的东西 the roof comb 房梁穿顶

commemorative [kəˈmemərətiv] *adj.* intended as a commemoration 纪念的(1)The new set of stamps will be issued as the commemo-rative. 这套新邮票将作为纪念物品发行。(2)The Royal Mint will strike a commemorative gold coin. 英国皇家造币厂将铸造一种纪念金币。

commercial [kəˈməːʃəl] *adj.* of, connected with, or engaged in commerce; mercantile; having profit as the main aim1.商业的，贸易的 (1)the commercial heart of the city 城市的商业中心(2)a commercial vehicle 商用车辆(3)commercial baby foods 市面上的婴儿食品 (4)the first commercial flights across the Atlantic 首次横跨大西洋的商业飞行 2.赢利的，以获利为目的的 3.偏重利润和声望的，商业化的

commission [kəˈmiʃən] *n.* the authority given to a person or organization to act as an agent to a principal in commercial transactions (权限,任务等的)委任,委托,(被委任的)任务 ,职权 1.（通常为政府管控或调查某

事的)委员会(1)The government has set up a commission of inquiry into the disturbances at the prison. 政府成立了一个委员会来调查监狱骚乱事件。(2) a commission on human rights 人权委员会 2.佣金，回扣(1)He gets a 10% commission on everything he sells.他从销售的每件商品中得到百分之十的佣金。(2)He earned £2000 in commission last month.他上个月挣了2000英镑的佣金。

compact ['kɔmpækt] *adj.* well constructed; solid; firm 紧凑的, 紧密的, 简洁的 1.袖珍的 a compact camera 袖珍照相机 2.紧凑的, 体积小的 The kitchen was compact but well equipped. 这间厨房虽然空间小但设备齐全。3.矮小而健壮的 He had a compact and muscular body. 他个子矮小身体健壮。

comply [kəm'plai] *v.* act in accordance with someone's rules, commands, or wishes 顺从, 答应 They refused to comply with the UN resolution. 他们拒绝遵守联合国的决议。

concrete ['kɔnkri:t] *n.* a strong hard building material composed of sand and gravel and cement and water 混凝土 These buildings are made of concrete and steel. 这些楼房是用钢筋和混凝土建成的。

consist of 由……组成 One week consists of seven days. 一周有七天。

conqueror ['kɔŋkərə] *n.* someone who is victorious by force of arms 征服者, 胜利者

conservancy [kən'sə:vənsi] *n.* conservation, esp. of natural resources 1.(港口、河流、地区等的)管理机构 the Thames Conservancy 泰晤士河管理委员会 Texas Nature Conservancy 得克萨斯州自然资源管理委员会 2.(对自然环境的)保护

conserve [kən'sə:v] *v.* keep in; safety and protect from harm, decay, loss, or destruction 保存, 保全

considerable [kən'sidərəbl] *adj.* large enough to reckon with; a lot of; much; worthy of respect 值得考虑的, 不可忽视的, 重要的, 相当(大, 多)的, 可观的(1)The project wasted a considerable amount of time and money.那项工程耗费了相当多的时间和资金。(2)Damage to the building was considerable. 这栋建筑物的损坏相当严重。

constitute ['kɔnstitju:t] *v.* (not in the continuous tenses 不用于进行时态) (fml 文) make up or form a whole; be the components of 组成, 构成(某整体), 为……之成分 We must redefine what constitutes a family. 我们必须重新定义家庭的概念。

construct [kən'strʌkt] *v.* to put together substances or parts,esp. systematically,in order to make or build (a building, bridge,etc)构成, 建造[筑], 施工 construct sth (from / out of / of sth)建筑, 修建, 建造(1)Mindful of the need to maintain efficient communication, the board of directors decided to construct a new communication network. 董事会注意到维持有效通讯的需要，因此决定建立一个新的通信网。(2)It took them two years to construct the bridge. 他们用了两年时间建这座桥。

construction [kən'strʌkʃən] *n.* the act or process of constructing; the art, trade, or work of building; the interpretation or explanation given to an expression or a statement 1.建筑, 建造, 施工(1)the construction industry 建筑业(2)road construction 道路的施工 (3)Work has been begun on the construction of the new airport. 新机场的修建已开工。(4)Our new office is still under construction. 我们的新办公楼尚在修建中。(5)the construction of a new database 新数据库的建立 2.建造(或构造)的方式 3.建筑物 The summer house was a simple wooden construction. 那座避暑别墅是简单的木结构建筑。4.(理念、观点和知识的)创立, 建立

consumption [kən'sʌmpʃən] *n.* (economics) the utilization of economic goods to satisfy needs manufacturing 消费 1.(能量、食物或材料的)消耗, 消耗量(1)Gas and oil consumption always increases in cold weather. 燃气和燃油的消耗量在天冷时总会增加。(2)The meat was declared unfit for human consumption. 这种肉被宣布不适于人食用。(3)He was advised to reduce his alcohol consumption. 他被劝告减少饮酒。(4)Her speech to party members was not intended for public consumption. 2. 消费 Consumption rather than saving has become the central feature of contemporary societies. 现代社会的主要特征是消费而不是储蓄。

contest ['kɔntest] *n.* a formal game or match in which two or more people, teams, etc, compete and attempt to win 1.比赛,竞赛(1) a boxing, archery, dancing, beauty, etc contest 拳击、射箭、舞蹈、选美等比赛(2)The election was so one-sided that it was really no contest, it only one side was likely to win. 选举呈现一边倒的局面,实际上毫无竞争可言。2.contest(for sth)争取获得控制权 *v.* 争论,争辩,竞赛,争夺

continent ['kɔntinənt] *n.* each of the main land masses of the Earth (Europe, Asia, Africa, etc)（地球上的）大洲之一（欧洲、亚洲、非洲等）,洲,大陆 2. the mainland of Europe 欧洲大陆,holidaying on the Continent 在欧洲大陆上度假

contour ['kɔntuə] *n.* outward curve of sth/sb (eg a coast, mountain range, body) thought of as defining its shape 轮廓,外形（如海岸、山脉、身体的轮廓线）(also contour line) line on a map joining points that are the same height above sea level 等高线 1.外形,轮廓 (1)The road follows the natural contours of the coastline. 这条路沿着海岸线的自然轮廓延伸。(2)She traced the contours of his face with her finger. 她用手指摸遍了他脸部的轮廓。2.(地图上表示相同海拔各点的)等高线 a contour map 等高线地图

contrast ['kɔntræst] *n.* a difference between two or more people or things that you can see clearly when they are compared or put close together 明显的差异,对比,对照 *v.* to compare two things in order to show the differences between them 对比,对照(1)Careful contrast of the two plans shows up some key differences. 把这两个计划仔细地加以对比就可以看出一些关键性的差异。(2)His white hair was in sharp contrast to (ie was very noticeably different from) his dark skin. 他的白头发与黑皮肤形成了鲜明的对比。(3)She had almost failed the exam, but her sister, by contrast, had done very well. 她考试差点不及格,而她的妹妹相比之下考得很好。

contribute [kən'tribju:t] *v.* to help bring about a result; act as a factor; to give or supply in common with others 贡献,捐献 (1)contribute aid for refugees 向难民提供援助。(2)Everyone should contribute what he or she can afford. 人人都应该尽自己的能力做贡献。(3)LuXun contributed greatly to Chinese literature. 鲁迅对中国文学做出了巨大的贡献。

controversial [ˌkɔntrə'və:ʃəl] *adj.* Marked by or capable of arousing controversy 争论的,引起争论的 (1)a highly controversial issue / topic / decision / book 颇有争议的问题/话题/决策/书 (2)a controversial plan to build a new road 有争议的筑路计划

convert...into... 把……转化成……

corbel ['kɔ:bəl] *n.* (architecture) a triangular bracket of brick or stone (usually of slight extent)承材,枕梁

cosmology [kɔz'mɔlədʒi] *n.* the metaphysical study of the origin and nature of the universe 宇宙学

couplet ['kʌplit] *n.* two items of the same kind;a stanza consisting of two successive lines of verse 对句[pl.]对联

cradle ['kreidl] *n.* a baby's bed with enclosed sides, often with a hood and rockers; a place where something originates or is nurtured during its early life 摇篮,发源地 1.摇篮 The mother rocked the baby to sleep in its cradle. 母亲摇动摇篮使婴儿入睡。2.cradle of sth 发源地,策源地,发祥地 Greece is the cradle of Western culture. 希腊是西方文化的发源地。

cross–fertilization [kɔrs- ˌfə:tilai'zeiʃən] *n.* interchange between different cultures or different ways of thinking that is mutually productive and beneficial 互育

curve [kə:v] *n.* line of which no part is straight and which changes direction without angles 曲线,弧线 *v.*(cause sth to) form a curve（使某物）成曲线形 1.曲线,弧线,面,弯曲 (1)the delicate curve of her ear 她耳朵的优美曲线(2)a pattern of straight lines and curves 直线与曲线交织的图案 (3)a curve in the road 道路上的拐弯

处(4)to plot a curve on a graph 在图上绘出一条曲线 2.(棒球运动中投向击球员的)曲线球

D

DC direct current 直流电

debate [di'beit] *n.* a discussion in which reasons are advanced for and against some proposition or proposal 辩论(1)A fierce debate on the tax cut was going on. 一场围绕着减税的辩论正在激烈进行中。(2)Her resignation caused much public debate. 她辞职一事引起群众议论纷纷。(3)After a long debate, the house of commons approved the bill. 经过长时间的辩论，下议院通过了议案。(4)After much debate, we decided to move to Oxford. 我们经充分讨论决定迁往牛津。(5)We had long debates at college about politics. 我们上大学时曾长时间地辩论政治问题。

decorative ['dekərətiv] *adj.* serving to decorate or adorn;ornamental 装饰的,装潢的,可作装饰的 (1)a large decorative candlestick having several arms or branches. 一个大烛台带有几个装饰臂和支架。(2) a decorative strip of cloth hung on a garment or wall.悬挂在衣服或墙上的装饰性布带 (3)The coloured lights are very decorative. 有这些彩灯大为生色。

decorate with 以……来装饰 People usually decorate the house with paper cuts.人们通常用剪纸装饰房子。

deliberate [di'libəreit] *adj.* by conscious design or purpose 1.故意的,有意的,蓄意的 a deliberate insult, lie, act 蓄意的侮辱,存心编造的谎言,故意的行为。2.不慌不忙的,小心翼翼的,审慎的(1) She has a slow, deliberate way of talking. 她谈话的方式缓慢而慎重。(2)making very deliberate gestures for emphasis. 做出非常沉稳的手势以示强调。

delicacy ['delikəsi] *n.* fine or subtle quality,character,construction,etc 娇嫩,优美精致(1)The Chinese palace lanterns are famous for their delicacy.中国的宫灯以精巧闻名。(2)the delicacy of her features 她那清秀的容貌。

deny [di'nai] *v.* to declare untrue; contradict; to refuse to believe; reject; to decline to grant or allow; refuse 否认,拒绝 1.否认,否定(某事)(1)He denied knowing anything about it. 他否认知道此事。2.拒不给与某人(所求或所需之物),阻止某人获得(所求或所需之物)(1)He gave to his friend what he denied to his family. 他宁可赠予朋友也不给家里人。(2)She was angry at being denied the opportunity to see me. 因不准她见我,她非常生气。

deriver from 起源,衍生,起源于 Many English words deriver from France. 很多英语单词源自法语。

descendant [di'send(ə)nt] *n.* a person, animal, or plant when described as descended from an individual, race, species, etc. 后代,子孙 1.后裔 (1)He was an O'Conor and a direct descendant of the last High King of Ireland. 他属于奥康纳家族,是爱尔兰最后一位君王的嫡系后裔。(2)Many of them are descendants of the original settlers. 他们中许多人都是早期移民的后裔。2.(由过去类似物发展来的)派生物

destroy [dis'trɔi] *v.* to ruin completely; spoil; to tear down or break up; demolish 1.摧毁,毁灭,破坏(1)The building was completely destroyed by fire. 这栋建筑被大火彻底焚毁了。(2)Heat gradually destroys vitamin C. 加热会逐渐破坏维生素C。2.(因动物有病或不再需要而)杀死,消灭,人道毁灭 The injured horse had to be destroyed. 这匹马受了伤,只好送它回老家了。

discipline ['disiplin] *n.* a branch of knowledge; a system of rules of conduct or method of practice 1.训练,纪律 (1)The school has a reputation for high standards of discipline. 这所学校因纪律严格而名闻遐迩。(2)Strict discipline is imposed on army recruits. 新兵受到严格的训练。2.训练方法,行为准则,符合准则的戒律 Yoga is a good discipline for learning to relax. 瑜伽是一种学会放松的有效方法。

discourse ['diskɔ:s] *n.*extended verbal expression in speech or writing 演讲,谈话,论述(1)I heard the professor's discourse on English lexicography. 我听了教授所作的关于英语辞典学的学术讲演。(2)They listened to his discourse on human relations. 他们听他作关于人际关系的演讲。(3)He discoursed impressively on Newton's theory of gravity. 他讲述了牛顿的引力定律,给人以深刻的印象。

dismantle [dis'mæntl] *v.* tear down so as to make flat with the ground; take apart into its constituent pieces 拆除……的设备,分解 1.拆开,拆卸(机器或结构)(1)I had to dismantle the engine in order to repair it. 我得把发动机拆开来修理。(2)The steel mill was dismantled piece by piece. 钢厂已经一块块拆散了。2.(逐渐)废除,取消 The government was in the process of dismantling the state-owned industries. 政府正在着手逐步废除国有企业。

display [di'splei] *vt.* to show or make visible 1.陈列,展出,展示(1)The exhibition gives local artists an opportunity to display their work. 这次展览为当地艺术家提供了展示自己作品的机会。(2)She displayed her bruises for all to see. 她将自己身上青一块紫一块的伤痕露出来给大家看。2.显示,显露,表现(特性或情感等)(1)I have rarely seen her display any sign of emotion. 我难得见到她将喜怒形于色。(2)These statistics display a definite trend. 这些统计数据表现出一种明显的趋势。

distinctly [dis'tiŋkt] *adv.* clear to the mind; with distinct mental discernment 清楚地,清晰地,无疑地,确实地 (1)I distinctly heard someone calling me. 我清楚地听到有人在叫我。(2)a distinctly Australian accent 明显的澳大利亚口音(3)He could remember everything very distinctly. 他什么事都能记得清清楚楚。

disturbing [di'stə:biŋ] *adj.* causing distress or worry or anxiety 令人不安的

diversified [dai'və:sifaid] *adj.* having variety of character or form or components; or having increased variety 多样化的 The culture has been diversified with the arrival of immigrants. 随着外来移民的到来,这里的文化变得多元化了。

dome [dəum] *n.* a concave shape whose distinguishing characteristic is that the concavity faces downward 圆屋顶 1.穹顶,圆屋顶 the dome of St Paul's Cathedral 圣保罗大教堂的穹顶 2.圆顶状物,穹状建筑物(1)his bald dome of a head 他那圆溜溜的秃顶(2)the Millennium Dome in London 伦敦的千年殿 3.(用于名称)圆顶体育场 the Houston Astrodome 休斯顿阿斯托洛圆顶运动场

dominant ['dɔminənt] *adj.* having primary control, authority, or influence; governing; ruling; predominant or primary 1.首要的,占支配地位的,占优势的,显著的(1)The firm has achieved a dominant position in the world market. 这家公司在国际市场上占有举足轻重的地位。(2)The dominant feature of the room was the large open fireplace. 房间的显著特色就是那巨大的明火壁炉。2.(基因)显性的,优势的

doorjamb ['dɔ:dʒæm] *n.* a jamb for a door 大门柱

drainage ['dreinidʒ] *n.* the action or a method of draining; a system of drains 1.排水,放水(1)a drainage system / channel / ditch 排水系统 / 渠 / 沟(2)The area has good natural drainage. 这个地区有良好的天然排水系统。2.排水系统

drill [dril] *n.* a tool with a sharp point and cutting edges for making holes in hard materials (usually rotating rapidly or by repeated blows)钻孔机,钻子,播种机 *v.* 1. to make a hole with a pointed power or hand tool 钻孔 2. to teach students, sports players etc by making them repeat the same lesson, exercise etc many times 训练,演练 She was drilling the class in the forms of the past tense. 她在训练学生做一般过去时语法练习。

due to 由于,归因于 The television station apologized for the interference, which was due to bad weather conditions. 电视台为出现的干扰表示歉意,那是由于恶劣的天气状况造成的。

dweller ['dwelə] *n.* a person who inhabits a particular place (1)a cave dweller 穴居者 (2)The noise of traffic is a constant irritant to city dwellers. 车辆的噪音对城市居民是永无止境的骚扰。

dynastic [dai'næstik] *adj.* of or relating to or characteristic of a dynasty 朝代的

dynasty ['dainəsti] *n.* a sequence of powerful leaders in the same family 朝代,王朝 the Nehru-Gandhi dynasty 尼赫鲁——甘地王朝

Dalai Lamas 达赖喇嘛

Dujiangyan Dam 都江堰

E

eclectic [ik'lektik] *adj.* selecting what seems best of various styles or ideas 不拘一格的,兼收并蓄的,折衷的 She has very eclectic tastes in literature. 她在文学方面的兴趣非常广泛。*n.* 折衷主义者,折衷派的人

Egypt ['i:dʒipt] *n.*埃及

Egyptian [i'dʒipʃən] *adj.* 埃及的,埃及人的

elaborate [i'læbərət] *adj.* marked by complexity and richness of detail; developed or executed with care and in minute detail 精致的,精巧的(1)elaborate designs / decorations 精心的设计/装饰(2)She had prepared a very elaborate meal. 她做了一顿精美的饭菜。(3)an elaborate computer system 精密的计算机系统

embody [im'bɔdi] *v.* to give a bodily form to; incarnate; to represent in bodily or material form 1.具体表现,体现,代表(思想或品质)(1)a politician who embodied the hopes of black youth 代表黑人青年希望的政治家(2)the principles embodied in the Declaration of Human Rights. 体现在《人权宣言》中的原则 2.包括,包含,收录 This model embodies many new features. 这种型号具有许多新特点。

embrasure [im'breiʒə] *n.* an opening (in a wall or ship or armored vehicle) for firing through 枪眼,【建】斜面墙

emission [i'miʃən] *n.*the act of emitting; causing to flow forth(光、热等的)发射,散发,喷射

emphasis ['emfəsis] *n.* special importance or significance; an object, idea, etc, that is given special importance or significance 强调,加强,侧重点,重要性,侧重或强调的事物 (1)There has been a shift of emphasis from manufacturing to service industries. 重点已经从制造业向服务行业转移。(2)The emphasis is very much on learning the spoken language. 重点主要放在学习口语上。(3)The course has a vocational emphasis. 这门课程着重职业培训。(4)We provide all types of information, with an emphasis on legal advice. 我们提供各种信息服务,尤其是法律咨询。(5)The examples we will look at have quite different emphases. 我们将要观察的例子所强调的重点很不相同。2.(对某个词或短语的)强调,加重语气,重读 "I can assure you," she added with emphasis, "the figures are correct." "我可以向你保证,"她加重语气补充道,"这些数字是正确的。"

emphasize ['emfəsaiz;'emfə,saiz] *v.*put emphasis on (sth.); give emphasis to (sth.); stress 强调,着重,加强(某词语)的语气,重读 We cannot emphasize too much the importance of learning English. 我们再怎样强调学英文的重要性也不为过。

enable...to do... 使……能做…… The loan enabled Jan to buy the house. 杰使用贷款买房子。

ensemble [ə:n'sɔ:mbl] *n.* a unit or group of complementary parts that contribute to a single effect; a coordinated outfit or costume 1.乐团,剧团,舞剧团(全体成员)(1)a brass / wind / string ensemble 铜管乐器/管乐器/弦乐器合奏组(2)The ensemble is / are based in Lyons. 这个乐团总部设在里昂。2.全体,整体 Her tout emsemble positively kills one. 她的外貌从整体上看确实使人着迷。

epitomize [i'pitəmaiz] *v.* to make an epitome of; sum up; to be a typical example of 摘要,概括,成为……缩影

essential [i'senʃəl] *n.*anything indispensable 要素,要点,必需品(1)In considering this problem, you should grasp its essentials. 在考虑这一问题时,你应当抓住实质。(2)Our course deals with the essentials of management. 我们的课程讲述管理的基本要点。(3)A knowledge of French is an absolute essential. 会些

法语是非常必要的。(4)We only had time to pack a few essentials. 我们只来得及装上几件必需品。

establish [is'tæbliʃ] *v.* to make secure or permanent in a certain place, condition, job, etc; to create or set up (an organization, etc) on or as if on a permanent 1.建立，创立，设立 The committee was established in 1912. 这个委员会创立与1912年。2.建立(尤指正式关系) The school has established a successful relationship with the local community. 这所学校与当地社区建立了良好的关系。3.establish sb / sth / yourself (in sth) (as sth) 确立，使立足，使稳固 4.获得接受，得到认可(1)It was this campaign that established the paper's reputation. 正是这场运动确立了这家报纸的声誉。(2)Traditions get established over time. 传统是随着时间的推移而得到认可的。

ethnic ['eθnik] *adj.* relating to or characteristic of a human group having racial, religious, linguistic, and certain other traits in common 种族的 1.民族的，种族的，部落的(1)ethnic grounds / communities 族群，种族社区(2) ethnic strife / tensions / violence 种族冲突 / 紧张局势 / 暴力(3)ethnic Albanians living in Germany 生活在德国的阿尔巴尼亚族人 2.具有民族特色的，异国风味的

exaggerate [ig'zædʒəreit] *v.* to enlarge beyond bounds or the truth 夸张，夸大其词(1)The hotel was really filthy and I'm not exaggerating. 我不是夸张，这旅店真的很脏。(2)He tends to exaggerate the difficulties. 他往往夸大困难。(3)I'm sure he exaggerates his Irish accent. 我肯定他的爱尔兰口音说得很重。

exchange [iks'tʃeindʒ] *v.* to give in return for something received; trade; interchange 1.交换，交流，掉换(1)to exchange ideas /news / information 交流思想，互通，交流信息(2)Juliet and David exchanged glances. 朱立叶和戴维相互看了看对方。(3)I shook hands and exchanged a few words with the manager. 我与经理握手，相互交谈了几句。2. exchange A for B 兑换，交易，更换(1)You can exchange your currency for dollars in the hotel. 你可在旅馆把你的钱兑换成美元。(2)If it doesn't fit, take it back and the store will exchange it.如果不合适就把它拿回来，商店将给你换掉。

exemplify [ig'zemplifai] *v.* clarify by giving an example of 1.是……的典型(1)Her early work is exemplified in her book, "A Study of Children's Minds." 她的《儿童思维研究》一书是她早期的代表作。(2)His food exemplifies Italian cooking at its best. 他的菜肴代表了意大利烹饪的最高峰。2.举例说明 She exemplified each of the points she was making with an amusing anecdote. 她的每一个论点都用一个逸闻趣事来说明。

exotic [ig'zɔtik] *adj.* having a strange or bizarre allure,beauty,or quality 式样奇特的，异国情调的(1)We saw pictures of exotic birds from the jungle of Brazil. 我们看到了来自于巴西热带雨林的各种奇异鸟类的照片。(2)Ignorance is like a delicate exotic fruit: touch it, and the bloom is gone.愚昧像个娇嫩奇异的水果，稍一触摸，就会失去其清新。

expand [iks'pænd] *v.* to increase the size, volume, quantity, or scope of; enlarge; to express at length or in detail; enlarge on 1.扩大，增加，增强(尺码、数量或重要性)(1)Metals expand when they are heated. 金属受热会膨胀。(2)Student numbers are expanding rapidly. 学生人数在迅速增加。(3)A child's vocabulary expands through reading. 孩子的词汇量通过阅读得到扩大。2.扩展，发展(业务)(1)We've expanded the business by opening two more stores. 我们增开了两个商店以扩展业务。(2)an expanding economy 不断发展的经济 3. expand on / upon sth 详细阐明 I repeated the question and waited for her to expand. 我把问题重复了一遍，等着她详细回答。

exposition [,ekspəu'ziʃən] *n.* a collection of things (goods or works of art etc.)for public display 博览会，展览会(1)They are going to hold an international exposition. 他们打算召开一次国际博览会。(2)an industrial exposition 工业博览会

exquisite ['ekskwizit] *adj.* extremely beautiful or delicate; finely or skilfully made or done 优美的，精致的(fml

文) (a) (of emotion) strongly felt; acute（指感情）感觉强烈的, 剧烈的 1.精美的, 精致的(1)exquisite craftsmanship 精美的工艺(2)Her wedding dress was absolutely exquisite. 她的婚纱真是漂亮极了。2.剧烈的, 强烈的 exquisite pain / pleasure 剧烈的疼痛 / 极大的快乐 3.微妙的, 敏锐的 (1)The room was decorated in exquisite taste. 这个房间的装饰情趣高雅。(2)an exquisite sense of timing 时间安排地恰到好处

exterior [eks'tiəriə] *n.* the region that is outside of something 1.外部, 外观, 表面, 外貌 The exterior of the house needs painting. 房子外墙需要油漆。2.（人的）外貌, 外表 Beneath his confident exterior, he was desperately nervous. 他表面上自信, 内心极度紧张。*adj.* situated in or suitable for the outdoors or outside of a building 外部的,外在的,表面的 The filming of the exterior scenes was done on the moors. 外景是在沼泽地拍摄的。

external [eks'tə:nl] *adj.* happening or arising or located outside or beyond some limits or esp. surface 1.外部的(1)the external walls of the building 建筑物的外墙(2)The lotion is for external use only. 此涂液仅限外用。2.外界的, 外来的, 在外的(1)A combination of internal and external factors caused the company to close down.内外因结合导致了公司的倒闭。(2)external pressures on the economy 外部因素对经济的压力3.外来的 An external auditor will verify the accounts. 外部审计员将核实这些项目。4.与外国有关的, 对外的(1)The government is committed to reducing the country's external debt. 政府决心减少本国的外债。(2)the Minister of State for External Affairs 外交大臣

extraordinarily [ik'strɔ:dθnərili] *adv.* extremely 非常(格外) (1)He behaves extraordinarily for someone in his position.对他那重地位的人来说, 他的行为很特别。(2)extraordinarily difficult 特别困难(3)She did extraordinarily well. 她干得特别好。

Edo period (1600-1868)江户时代
Emperor Takaaki 孝明天皇
Emperor Yangdi of Sui Dynasty 隋炀帝
Eclectic Era【建】折衷主义建筑时代（十九世纪上半叶至二十世纪初, 在欧美一些国家流行的一种建筑风格。折衷主义建筑师任意模仿历史上各种建筑风格, 他们不讲求固定的方式, 只讲求比例均衡, 注重纯形式美。）
Eight wonders in the world 世界八大奇迹

F

fabled ['feibld] *adj.* celebrated in fable or legend 虚构的(1)a fabled monster 传说中的巨怪(2)For the first week he never actually saw the fabled Jack. 第一周他实际上从没见到传奇式的杰克。

facade [fə'sɑ:d] *n.* the face or front of a building 建筑物的正面 1.a classical facade 古典式建筑正面2.(虚假的)表面, 外表(1)She managed to maintain a facade of indifference. 她设法继续装作漠不关心的样子。(2)Squalor and poverty lay behind the city's glittering facade. 表面的繁华掩盖了这座城市的肮脏和贫穷。

facilitate [fə'siliteit] *v.* make easier 促进, 帮助, 使容易 Equipping an office or plant with computers to facilitate or automate procedures. 给办公室或工厂装备计算机, 使办公手续或生产过程更方便或自动化。

far from 远离, 远远不, 完全不, 非但不 He traveled far away from home. They sat far away from each other. 他长途跋涉, 远离家乡。他们彼此远远地坐着。

feature ['fi:tʃə] *n.* any of the distinct parts of the face, as the eyes, nose, or mouth 1.特色, 特征, 特点 The software has no particular distinguishing features. 这个软件没有明显的特征。geographical features 地势

2.面容的一部分(如鼻、口、眼)(1)his strong handsome features 他轮廓分明的英俊面孔。(2)Her eyes are her most striking feature. 她容貌中最引人注目的是她的双眼。3.feature (on sb / sth) (报章、电视等的)特写,专题节目 a special feature on education 关于教育的专题文章 4.(电影的)正片,故事片 v. to give special attention to; display, publicize, or make prominent; 以……为特色,以……为号召 1. feature sb / sth (as sb / sth)以……为特色,是……的特征 Many of the hotels featured in the brochure offer special deals for weekend breaks. 小册子例举的多家旅馆都有周末假日特别优待。2. feature (in sth)起重要作用,占重要地位 Olive oil and garlic feature prominently in his recipes. 橄榄油和大蒜在他的食谱中显得很重要。

Fengshui 风水

ferment ['fə:ment] *n.* commotion; unrest 激动,纷扰 The country was in ferment. 那个国家处于动乱中。

Flamboyant [flæm'bɔiənt] *adj.*【建】火焰式的

flank [flæŋk] *n.* the side of building 侧面,【建】厢房

filter ['filtə] *v.* to pass (through a filter or something like a filter) 过滤,渗透 (1)The water was filtered through charcoal. 这水是用木炭来过滤的。(2)News of the defeat filtered through. 失败的消息走漏了出来。(3) Sunlight filtered through the leaves. 阳光透过树叶射下来。(4)You need to filter the drinking water. 饮用水需要过滤。(5)The sunlight filtered through the curtains. 阳光透过窗帘映了进来。

flavor ['fleivə] *n.* the general atmosphere of a place or situation and the effect that it has on people; the taste experience when a savoury condiment is taken into the mouth 1.味道(1)The tomatoes give extra flavor to the sauce. 番茄使调味汁别有风味。(2)It is stronger in flavor than other traditional Dutch cheeses. 这比传统荷兰干酪味道要浓。2.特色,气氛(1)the distinctive flavor of South Florida 南佛罗里达的独特风情(2) Foreign visitors help to give a truly international flavour to the occasion. 外国客人使这个场合显出一种真正国际性的气氛。

flexible ['fleksəbl] *adj.* capable of being bent or flexed; pliable 1.能适应新情况的,灵活的,可变动的(1)a more flexible design /approach 更灵活的设计/方法(2)flexible working hours 弹性工作时间(3)Our plans need to be flexible enough to cater for the needs of everyone. 我们的计划必须能够变通,以满足每个人的需要。(4)You need to be more flexible and imaginative in your approach. 你的方法必须更加灵活,更富有想象力。2.柔韧的,可弯曲的,有弹性的 flexible plastic tubing 挠性塑料管

flourish ['flʌriʃ] *v.* very active, or widespread; prosper 繁荣,兴旺,活跃,盛行(1)This type of plant flourishes in hot countries. 这种植物在热带国家生长茂盛。(2)There were two or three palm trees flourishing in the promenade garden. 街心花园里有两三棵枝繁叶茂的棕榈树。(3)All the family are flourishing. 全家人身体都很好.(4)His business is flourishing. 他的生意兴隆。(5)No new business can flourish in the present economic climate. 在目前的经济气候中,任何生意都兴旺不起来。(6)Art flourished in that period. 那个时期艺术十分繁荣。(7)In Germany the baroque style of art flourished in the 17th and 18th centuries. 在德国,巴罗克艺术风格在17和18世纪非常盛行。

focus on 集中 Focus your attention on your work. 把注意力集中在你的工作上。

foray ['fɔrei] *n.* brief but vigorous attempt to be involved in a different activity(对一新的活动)短暂而积极的尝试(1)An actor's foray into politics. 一演员对政治的尝试性介入(2)go on/make a foray into enemy territory 袭击敌占区

forge [fɔ:dʒ] *v./n.* to put a lot of effort into making sth.successful or strong so that it will last 艰难干成,努力加强,to move forward in a steady but powerful way 稳步前进(1)Donald was forced to forge a signature. 唐纳德被迫伪造签字。(2)The ship forged ahead under a favorable wind. 船乘风快速前进。(3)They forged their manager's signature on the cheque. 他们在支票上伪造了经理的签名。

forum ['fɔ:rəm] *n.* a public meeting or assembly for open discussion 论坛,讨论会(1)They are holding a forum on juvenile delinquency. 他们正举行一个有关青少年犯罪的讨论会。(2)The letters' page of this newspaper is a forum for public argument. 这份报纸的读者来信栏是公众意见的论坛。(3)The letters' page serves as a useful forum for the exchange of readers' views. 读者来信版是读者们交换意见的园地。

frustrate [frʌs'treit] *vt.* to hinder or prevent (the efforts, plans, or desires) of; thwart; to upset, agitate, or tire 1.使懊丧,使懊恼,使沮丧 What frustrates him is that there's too little money to spend on the project.使他懊恼的是可用于这个项目的资金太少。2.阻止,防止,挫败 The rescue attempt was frustrated by bad weather. 营救行动因天气恶劣受阻。

Functionalism ['fʌŋkʃənəlizəm] 【建】功能主义建筑（认为建筑形式应该服从功能,主张建筑或物品设计或风格首要的是用途而不是外观）

fundamentally [fʌndə'mentəl;] *adv.* in a way that affects the basis or essentials; utterly 1.根本上,完全地(1)The two approaches are fundamentally different. 这两个处理方法完全不同。(2)By the 1960s the situation had changed fundamentally. 到20世纪60年代形势已发生了根本的变化。(3)They remained fundamentally opposed to the plan. 他们依然从根本上反对这项计划。2.（引入话题时说）从根本上说,基本上 Foundamentally, there are two different approaches to the problem. 从根本上说,这个问题有两种不同的处理方法。3.（表示最重要的方面）根本上,基本上 She is fundamentally a nice person, but she finds it difficult to communicate. 她基本上是个好人,但她觉得难以和人沟通。

folk residences 民居

Forbidden City 紫禁城

foundationally speaking 从根本上来说 Foundationally speaking,the project has been cancelled because of lack of funds. 该项目已经被取消的根本原因是缺少资金。

fight for 为……而战 They fight for defend the country. 他们为保卫祖国而战斗。

fill with 充满,用……填充 Fill the bottle with water. 把瓶子装满水。

G

gargoyle ['gɑ:gɔil] *n.* stone or metal spout in the form of a grotesque human or animal figure, for carrying rain-water away from the roof of a church, etc 滴水嘴,(疏导雨水的)凸饰漏嘴,怪形雕塑的放水口

garrison ['gærisn] *n.* troops stationed in a town or fort 卫戍部队,守备部队,警备部队(1)a garrison of 5,000 troops 有5,000士兵驻守的防地(2)a garrison town 有驻防的城市(3)Half the garrison is / are on duty. 卫戍部队有半数在执行守备任务。*v.* ~ sth (with sb) defend (a place) with or as a garrison 卫戍部队守备（某地）(1)Two regiments were sent to garrison the town. 派了两个团驻守在那个城镇。(2)100 soldiers were garrisoned in the town. 派了100名士兵在城里驻防。

gear [giə] *n.* equipment, such as tools or clothing, used for a particular activity 1.排挡,齿轮,传动装置 Careless use of the clutch may damage the gears. 离合器使用不慎可能会损坏传动装置。2.挡 reverse gear 倒挡 low / high gear 低速/高速挡 to change gear /to shift gear 换挡 3.（某种活动的）设备,用具,衣服 climbing / fishing / sports gear 爬山/钓鱼/运动用具 4.衣服 wearing the latest gear 穿着最新款式的衣服 5.（特定用途的）器械,装置 lifting / towing / winding gear 起重/拖拽/卷扬装置, move into high gear 进入高速发展 get into gear / get sth into gear （使）开始工作 out of gear 失去控制

generator ['dʒenəreitə] *n.* one that generates, esp. a machine that converts mechanical energy into electrical energy. 发电机,生产者 1.发电机(1)The factory's emergency generators were used during the power cut. 工厂应急

发电机在停电期间用上了。(2)a wind generator 风力发电机 2.发生器(1)The museum uses smells and smoke genarators to create atmosphere. 博物馆利用气味和烟雾发生器制造气氛。(2)The company is a major generator of jobs. 这家公司创造了相当多的就业机会。3.电力公司 the UK's major electricity generator 英国主要的电力公司

generally speaking 一般而言 Generally speaking, I think you're right. 一般而言,我认为你是对的。

giant ['dʒaiənt] *adj.* marked by exceptionally great size, magnitude, or power 庞大的, 巨大的(1)a giant crab 巨蟹(2)a giant-size box of tissues 一大盒纸巾(3)a giant step towards achieving independence 朝着独立迈出的巨大的一步 *n.* a person or thing of great size 巨大的动物或植物, 伟人, 天才 1.(故事中常为惭愧而愚蠢的)巨人 He's a giant man. 他是个巨人。2.大公司, 强大的组织 the multinational oil giants 跨国大石油公司 3.伟人, 卓越人 literary giant 大文豪

gigantic [dʒai'gæntik] *adj.* of very great size or extent; immense 巨大的, 庞大的

Giza ['gi:zə] *n.* 吉萨省(埃及省份,位于开罗西南面)

glaze [gleiz] *v./n.* to cover something with a glaze to give it a shiny surface 给……上釉, 使光滑, 使光亮 to fit sheets of glass into something 给……安装玻璃 to thin clear liquid put on clay objects such as cups and plates before they are finished,to give them a hard shiny surface 釉, 釉料 (1)Glaze the pie with beaten egg. 在馅饼上涂上打匀的蛋液使表面发亮。(2) I know how to glaze a window. 我知道怎么给窗户装玻璃。(3) They found several glazed clay pots in the ancient tomb. 他们在古墓中发现了一些上釉的陶罐。

glutinous ['glu:tinəs] *adj.* resembling glue in texture; sticky 粘的, 粘质的 glutinous rice 糯米

Gothic ['gɔθik] *adj.* 【建】哥特式的(尖拱式建筑)

grand [grænd] *adj.* large and impressive in size, scope, or extent; magnificent 1.壮丽的, 堂皇的 (1)It's not a very grand house. 这房子并不是十分富丽堂皇。(2)The wedding was very grand occasion. 婚礼场面非常隆重。2.(用于大建筑物等的名称)大(1)the Grand Canyon 大峡谷(2)We stayed at the Grand Hotel. 我们住在格兰酒店。3.宏大的, 宏伟的, 有气派的(1)a grand design / plan / strategy 宏伟的蓝图 / 宏大的计划 / 重大的战略思想(2)New Yorkers built their city on a grand scale. 纽约人大规模地建造自己的城市。(对上层社会的人的称呼)大(1)the Grand Duchess Elena 大公夫人埃琳娜(2)He described himself grandly as a "landscape architect". 他自封为"景观建筑师"。

grandeur ['grændʒə] *n.* the quality or condition of being grand; magnificence 庄严, 伟大 1.宏伟, 壮丽, 堂皇(1) the grandeur and simplicity of Roman architecture 古罗马建筑的雄伟和简朴(2)The hotel had an air of faded grandeur. 这饭店给人一种繁华已逝的感觉。2.高贵, 显赫, 伟大(1)He has a sense of grandeur about him. 他觉得自己很了不起。(2)He is clearly suffering from delusions of grandur. 他显然是犯了妄自尊大的毛病。

granite ['grænit] *n.* a very hard grey rock, often used in building 花岗石

grave [greiv] *adj.* dignified and somber in manner or character and committed to keeping promises 1.严重的, 重大的, 严峻的, 深切的(1)The police have expressed grave concern about the missing child's safety. 警方对失踪孩子的安全深表关注。(2)The consequences will be very grave if nothing is done. 如果不采取任何措施后果将会是非常严重的。(3)We were in grave danger. 我们处于极大的危险之中。2.严肃的, 庄严的, 表情沉重的 He looked very grave as he enteredthe room. 他进屋时表情非常严肃。*n.* a place for the burial of a corpse (esp. beneath the ground and marked by a tombstone)墓 turn in his / her grave 九泉之下不得安宁 My father would turn in grave if he knew. 我父亲知道的话, 他在九泉之下也不会安宁。

guide [gaid] *v.* to serve as a guide for; conduct; to direct the course of; steer 带领, 为……做向导, 带领, 指引 1.guide sb (to / through / around)给某人领路(或导游), 指引(1)She guided us through the busy streets to the cathedral. 她带领我们穿过繁忙的街道去大教堂。(2)We were guided around the museums. 我们在导游的带领下参观了博物馆。2.指导，影响（某人的行为）He was always guided by his religious beliefs. 他的言行总被他的宗教信仰所支配。3. guide sb (through sth)（向某人）解释，阐明 The health and safety officer will guide you through the safety procedures. 健康安全官员将给你解释一遍安全规程。

gutter ['gʌtə] *n.* long (usu semicircular) metal or plastic channel fixed under the edge of a roof to carry away rain-water 排水檐沟，天沟(channel at the) side of a road, next to the kerb 路边沟，排水沟，阴沟

Great Wall at Mutianyu 慕田峪长城

Getty Center 洛杉矶盖蒂中心（包括一座非常现代化的美术博物馆、一个艺术研究中心和一所漂亮的花园）

H

handrail ['hændreil] *n.* a narrow railing to be grasped with the hand for support 栏杆，扶手

harmonious [hɑ:'məunjəs] *adj.* exhibiting accord in feeling or action; having component elements pleasingly or appropriately combined 和谐的，协调的 a harmonious combination of colours 协调的色彩搭配

harmony ['hɑ:məni] *n.* agreement in action, opinion, feeling, etc; accord 1.融洽，和睦(1)the need to be in harmony with our environment 同我们的环境协调的必要(2)to live together in perfect harmony 十分和睦地生活在一起(3)social / racial harmony 社会／种族融洽 2.和声(1)a sing in harmony 用和声唱(2)to study four-part harmony 学习四部和声(3)passionate lyrics and stuning vocal harmonies 充满激情的歌词和绝妙的和声演唱 3.和谐 the harmony of colour in nature 自然界色彩的协调

headquarters ['hed,kwɔ:təz] *n.* any centre or building from which operations are directed, as in the military, the police, etc (公司,机关等的)总部，总公司，总局，(军,警的)司令部，总部，总署，司令部(全体指挥)人员(1)The firm's headquarters is / are in London. 公司总部设在伦敦。(2)Several companies have their headquarters in the area. 有几家公司总部设在这个地区。(3)I'm now based at headquarters. 我现在在总公司工作。(4)police headquarters 警察总局(5)Headquarters in Dublin has / have agreed. 都柏林总部都已经同意了。

heel [hi:l] *n.* the back part of the human foot 脚后跟(1)I have a blister on my heel because my shoe is too tight. 鞋子太紧了,我脚后跟起了个泡。(2)There is a hole in the heel of one of your socks. 你一只袜子的后跟有个洞。(3)She prefers wearing high heels on formal occasions. 在正式场合她比较喜欢穿高跟鞋。

Heian-kyo 平安京

Heian Shrine 平安神社

hostility [hɔs'tiliti] *n.* fighting; warfare 反对，抵抗[pl.]战争（状态），战斗(1)We had no hostility toward the new neighbor. 我们对新邻居毫无敌意。(2)Hostilities ended when the treaty was signed. 条约签订后战争结束了。(3)After weeks of silent hostility they've at last had it out with each other. 他们经过几个星期的暗斗之后,彼此终于谈开了。(4)Humour was his only weapon against their hostility. 他有幽默感,这是他对付敌对行动的唯一手段。(5)Before getting the proposals accepted, the government had to run the gauntlet of hostility from its own supporters. 拥护政府的人对政府进行了尖锐的抨击之后,这些建议才得以接受。(6)His suggestion met with some hostility. 他的建议遭到某种程度的反对。

Huizhou residences 徽州民居

Hakka earth buildings in Fujian 福建土楼

have very little to do with 与……没有多大关系

I

iconographic [ai,kɔnə'græfik] *adj.* of iconography 肖像的, 肖像学的, 肖像画法

identity [ai'dentiti] *n.* the distinct personality of an individual regarded as a persisting entity 1.身份 (1)The police are trying to discover the identity of the killer. 警方正努力调查杀人凶手的身份。(2)Their identities were kept secret. 他们的身份保密。(3)She is innocent; it was a case of mistaken identity. 她是无辜的, 那是身份判断错误。(4)Do you have any proof of identity? 你有身份证明吗？ 2.特征, 特有的感觉(或信仰)a sense of national / cultural / personal / group identity 民族 / 文化 / 个人 / 群体特性的认同感 3.identity (with sb / sth), identity (between A and B)同一性, 一致 There's a close identity between fans and their team. 球迷和他们的球队之间有密切的同一性。

illustrate ['iləstreit] *v.* to clarify, as by use of examples or comparisons; to provide (a publication) with explanatory or decorative features 举例说明, 图解, 加插图于, 阐明 1.illustrate sth (with sth)加插图于, 给(书等)做图表(1) an illustrated textbook 有插图的课本(2)His lecture was illustrated with slides taken during the expedition.他在讲演中使用了探险时拍摄的幻灯片。2.(用示例、图画等)说明, 解释(1)To illustrate my point, let me tell you a little story. 为了说明我的观点, 让我来给你们讲个小故事。(2)Last year's sales figures are illustrated in Figure 2. 图 2 显示了去年的销售数字。3.表明……真实, 显示……存在 The incident illustrates the need for better security measures. 这次事件说明了加强安全措施的必要。

imitate ['imiteit] *v.* reproduce someone's behavior or looks; appear like, as in behavior or appearance 1.模仿, 仿效(1)Her style of painting has been imitated by other artists. 她的绘画风格为其他画家所模仿。(2)Art imitates Nature. 艺术是对大自然的仿制。(3)Teachers provide a model for children to imitate. 教师是孩子们仿效的典范。(4)No computer can imitate the complex functions of the human brain.任何计算机都无法模拟人脑的复杂功能。2.模仿(某人的讲话、举止), 作滑稽模仿 She knew that the girls used to imitate her and laugh at her behind her back. 她知道女孩子们过去常在背地里模仿她、嘲笑她。

imperial [im'piəriəl] *adj.* of, relating to, or suggestive of an empire or a sovereign, esp. an emperor or empress 1.帝国的, 皇帝的 (1)the imperial family / palace / army 皇室家族, 皇宫, 皇家陆军(2)imperial power / expansion 皇权, 帝国的扩张 2.(度量衡)英制的

imposing [im'pəuziŋ] *adj.* impressive in appearance or manner; grand (外表或举止)壮观的, 令人印象深刻的(1) a grand and imposing building 雄伟壮观的建筑物 (2)a tall imposing woman 高大壮硕的女人

impression [im'preʃən] *n.* a vague idea, consciousness, or belief 印象(1)The new teacher made a good impression on the students. 新教师给学生留下了一个好印象。(2)Her gentleness has given me a deep impression.她的亲切给我留下了很深的印象。(3)The robber left an impression of his feet in the mud. 强盗在烂泥里留下了他的脚印。(4)My advice seemed to make little impression on him. 我的忠告似乎对他不起作用。(5) Punishment seemed to make no impression on the child. 惩罚对这孩子似乎没什么效果。(6)I have the impression that I've seen that man before. 我觉得我以前见过那个人。(7) The students did some marvellous impressions of the teachers at the end-of-term party. 在期末联欢会上, 学生模仿教师惟妙惟肖, 令人捧腹。

inaccessible [,inæk'sesəbl] *adj.* capable of being reached only with great difficulty or not at all 难接近的,难达到的(1)They live in a remote area, inaccessible except by car. 他们住在一处偏远地区, 只能开车去。(2)Dirt can collect in inaccessible places. 尘土会积聚在人够不着的地方。(3)The temple is now inaccessible to the public. 这个寺庙现在不对公众开放了。(4)The language of teenagers is often completely inaccessible to adults.成人往往完全听不懂青少年的语言。

in accordance with 与……一致,依照 He acted in accordance with his beliefs. 他依照自己的信念行事。

in a sense 从某种意义上说 In a sense, your personality lies in your sense of humor.

in a state of 处于……状态 In a state of extreme emotion 处于极度热情状态下的

in contrast to/with 和……形成对比 His white hair was in sharp contrast to his dark skin. 他的白发和他的黑皮肤形成鲜明的对比。

incorporate ['inkɔpərit] *v.* to unite (one thing) with something else already in existence; to cause to form into a legal corporation 合并,使组成公司,具体表现 incorporate sth (in / into/ with sth) 将……包括在内,包含,吸收,使并入(1)Many of your suggestions have been incorporated in the plan. 你的很多建议已纳入计划中。(2)The new car design incorporates all the latest safety features. 新的汽车设计包括了所有最新的安全配备。(3)We have incorporated all the latest safety features into the design. 我们在设计中纳入了所有最新的安全装置。

incredible [in'kredəbl] *adj.* beyond belief or understanding; unbelievable; marvellous; amazing 1.难以置信的(1) an incredible story 不可思议的故事(2)It seemed incredible that she had been there a week already. 真正让人难以相信,她已经在那里待了一个星期了。2.极好的,极大的(1)The hotel was incredible. 这家旅馆棒极了。(2)an incredible amount of work 量极大的工作

individual [,indi'vidjuəl] *adj.* being or characteristic of a single thing or person 个人的,个体的,个别的,单独的(1)The director of the factory felt no individual responsibility for the deficit. 厂长觉得工厂亏损没有任何个人责任。(2)Each individual person is responsible for his own arrangements. 每人均须对自己的计划负责。(3)Students can apply for individual tuition. 学生可以申请个别授课。(4)It is difficult for a teacher to give individual attention to children in a large class. 教师在人数多的班上很难对各个学生都照顾到。(5) The model has an individual way of dressing. 这个模特儿有着独特的穿衣方式。

influence on 对……有影响 This book has great influence on readers. 这本书对读者产生了长远影响。

influential [,influ'enʃəl] *adj.* having or exerting influence 有影响的,有权势的(1)a highly influential book 十分有影响力的书(2)She is one of the most influential figures in local politics. 她是本地政坛举足轻重的人物。(3)The committee was influential in formulating government policy on employment. 委员会左右着政府就业政策的制定。

initial [i'niʃəl] *adj.* of, at, or concerning the beginning 最初的,开始的,初期的(1)an initial payment of £60 and ten instalments of £25*60 英镑的首期付款加十次25英镑的分期付款(2)in the initial stages of the campaign 运动的最初阶段(3)My initial reaction was to decline the offer. 我最初的反应是婉言谢绝这个提议。

initiate [i'niʃieit] *n.* someone new to a field or activity 入会,开始,新加入某组织(或机构、宗教)的人,新入会的人 *v.* 开始,创始,启蒙 bring into being; take the lead or initiative in; participate in the development of (1)to initiate legal action / proceedings against sb 起诉某人(2)The government has initiated a programme of economic reform. 政府已开始实施经济改革方案。

innovation [,inəu'veiʃən] *n.* a creation (a new device or process)resulting from study and experimentation 创新,革新(1)A surge of innovation in techniques is on the horizon.技术改进的浪潮即将出现。(2)Anesthesia was a great innovation in medicine. 麻醉是一项伟大的医学创新。(3)The innovation of air travel during this century has made the world seem smaller. 本世纪空中旅行的革新使世界似乎变小了。(4)technical innovations in industry 工业技术革新

inspire [in'spaiə] *v.* to arouse,to prompt or instigate 鼓舞,激起,启发 (1)His speech inspired us to try again. 他的演讲鼓舞了我们再作尝试。(2)The beautiful scenery inspired the composer. 美丽的景色使作曲家灵思泉

涌。(3)His encouraging remarks inspired confidence in me. 他的一番鼓励激起了我的信心。(4)The riot was inspired by extremists. 暴乱是由极端主义者鼓动的。(5)The painting can inspire a pensive mood.这幅画能引人沉思。(6)The sight inspired him with nostalgia. 这景象激起了他的怀旧之情。(7)His best music was inspired by the memory of his mother. 他最好的乐曲创作灵感来自对母亲的怀念。

instrument ['instrumənt] *n.* a device that requires skill for proper use 仪器，器械，工具，手段，乐器(1)This instrument monitors the patient's heartbeats.这台仪器监听病人的心跳。(2)The compass is an instrument of navigation 罗盘是航行仪器。(3)The viola is a stringed instrument. 中提琴是一种弦乐器。(4)Language is an instrument for communication. 语言是交际的手段。(5)The organization he had built up eventually became the instrument of his downfall. 他创建起来的组织到头来却成为促使他倒台的根本原因了。(6)The king signed the instrument of abdication. 国王签署了逊位的文告。

integrate ['intigreit] *v.* to make into a whole by bringing all parts together; unify.To join with something else; unite...使……结合(with),使并入(into),使一体化,使完全,使成一整体 1.integrate (A) (into / with B) ; integrate A and B (使)合并,成为一体(1)These programs will integrate with your existing software. 这些程序将和你已有的软件整合成一体。(2)These programs can be integrated with your existing software. 这些程序能和你已有的软件整合成一体。2.integrate (sb) (into / with sth)(使)加入,融入群体 They have not made any effort to integrate with the local community.他们完全没有尝试融入本地社区。

intellectual [,intəl'ektʃuəl] *adj.*of or associated with or requiring the use of the mind 智力的，聪明的，理智的(1)intellectual powers 智力(2)intellectual people 善思考的人(3)Chess is a highly intellectual game. 象棋是需用高度智力的运动项目。(4)He likes to set himself up as an intellectual. 他喜欢自命为知识分子。(5)We admired his intellectual providence to acquire vast stores of dry information. 我们钦佩他收集大量原始资料的远见卓识。

in terms of 根据,按照,就……而言 Our firm is very strong in terms of manpower. 我们的公司在人力资源方面非常强大。

interruption [,intə'rʌpʃən] *n.* something that interrupts; an interval or intermission 中断，打断，障碍物，遮断物(1)The rain continued without interruption all day. 这雨整天下个不停。(2)Constant interruptions interfere with my work. 一次又一次的干扰妨碍我的工作。(3)This interruption is very annoying. 这种打扰真讨厌。(4)My speech went quite well until I was put off my stroke by the interruption. 起初我讲得很顺利,可是受到干扰后就结巴起来了。(5)He was impatient of any interruption. 他对任何打扰都感到不耐烦。

intellectual [,inti'lektjuəl] *n.* a person who uses the mind creatively 知识分子,有知识者

invasion [in'veiʃən] *n.* the act of invading; any entry into an area not previously occupied 1.武装入侵,侵略,侵犯(1)the German invasion of Poland in September,1939 德国1939年9月对波兰的入侵(2)the threat of invasion 入侵的威胁(3)an invasion force / fleet 侵略军 / 舰队 2.(尤指烦恼的)涌入(1)the annual tourist invasion 一年一度游客的涌入(2)Farmers are struggling to cope with an invasion of slugs. 农民正在努力对付蛞蝓的大肆侵害。3.侵犯,干预 The actress described the photographs of her as invasion of privacy. 那位女演员认为她的这些照片是对隐私权的侵犯。

investigation [in,vesti'geiʃən] *n.* a careful search or examination in order to discover facts,etc 调查,调查研究(1)It's only the initiative of the investigation.这仅仅是调查的开始。(2)We were greatly aided in our investigation by the cooperation of the police.我们在调查的过程中得到警方的大力协助。(3)The investigation was carried out under the direction of a senior police officer. 调查是在一位高级警官的指导下进行的。(4)The investigation into the accident was carried out by two policemen. 两名警察对这一事故展开调查。

irrigate ['irigeit] *v.* to supply (dry land) with water by means of ditches, pipes, or streams; to water artificially; to wash out (a body cavity or wound) with water or a medicated fluid (1)灌溉 irrigated land / crops经过灌溉的土地 / 农作物 (2)冲洗(伤口或身体部位)

irrigation [,iri'geiʃən] *n.* supplying dry land with water by means of ditches etc 灌溉, 冲洗

ICT Information and Communication Technology 信息和通信技术

Ieoh Ming Pei 贝聿铭(1917-), 著名华裔美国建筑大师

in full swing 正在全力进行中 The building project is in full swing. 这项建筑工程正在全力进行中。

J

joint [dʒɔint] *n.* the point of connection between two bones or elements of a skeleton 木模, 接榫(1)He suffers from arthritis in his leg joints. 他的腿有关节炎。(2)The pipe was rusted at the joint. 管子的接合处生锈了。(3)Take a seat, have a cigarette or a joint and I will be back in three minutes. 请坐,抽根香烟,我三分钟内就回来。

joist [dʒɔist] *n.* one of the long thick pieces of wood or metal that are used to support a floor or ceiling in a building 搁栅, 托梁

jungle ['dʒʌŋgl] *n.* an impenetrable equatorial forest 1.(热带)丛林, 密林(1)The area was covered in dense jungle. 这个地区丛林密布。(2)the jungles of South-East Asia 东南亚热带丛林 (3)jungle warfare 丛林战 (4)Our garden is a complete jungle. 我们的花园杂草丛生。2.尔虞我诈的环境, 危险地带 It's a jungle out there—you've got to be strong to succeed. 那是个弱肉强食的地方, 要成功就得是强者。

K

keystone ['ki:stəun] *n.* a central cohesive source of support and stability; the central building block at the top of an arch or vault 拱心石, 楔石, 重点

Khmer [kə'mɛə] *n.* the Mon-Khmer language spoken in Cambodia; a native or inhabitant of Cambodia 谷美尔人, 谷美尔语

Khufu ['ku:fu:] *n.* 胡夫(古埃及第四王朝法老, 在位期间下令修建了最大的金字塔)

Kunming Lake 昆明湖

Kyoto 古京都

Kyoto Gosho 京都御所

Kenzo Tange 丹下健三, 世界著名的日本建筑师, 在五十年代, 赢得了几乎每一个他参加的竞赛, 完成了一系列雄伟的公共建筑与国家事务核心设施等大型设计案

Konarka Power Plastic 科纳卡光电塑料技术

L

landmark ['lændmɑ:k] *n.* a prominent or well-known object in or feature of a particular landscape; an important or unique decision, event, fact, discovery, etc (显而易见的)地标, 陆标, (历史上划时代的)重大事件, 里程碑, 路标, 地标(有助于识别所处地点的大建筑物)The Empire State Building is a familiar landmark on the New York skyline. 帝国大厦是人们熟悉的纽约高楼大厦中的地面标志物。

landscape ['lændskeip] *n.* an expanse of scenery that can be seen in a single view 风景, 山水画, 地形, 前景1.(陆

上,尤指乡村的)风景,景色(1)the rugged / mountainous / dramatic landscape of Bolivia 玻利维亚崎岖的陆地景观 / 重峦叠嶂的地貌 / 给人深刻印象的风光(2)the woods and fields that are typical features of the English landscape 具有典型英国风景特征的森林与田野(3)an urban / industrial landscape 都市 / 工业景观(4)We can expect changes in the political landscape. 我们等着看政治舞台上的变化吧。2.乡村风景画,乡村风景画的风格 a British artist famous for his landscapes 以风景画而闻名的英国画家 3.(文件的)横向打印格式 Select the landscape option when printing the file. 打印文件时应选择横向打印格式选项。

lime [laim] *n./v.* a white substance obtained by heating limestone,used in building materials and to help plants grow 石灰 to add the substance lime to soil,esp. in order to control the acid in it(尤指为控制酸度而给土壤)掺加石灰,撒石灰

limestone ['laimstəun] *n.* a sedimentary rock consisting mainly of calcium that was deposited by the remains of marine animals 石灰石 Both companies locate at the neighborhood of our factory. 这两家公司都位于我们工厂邻边。

lintel ['lintl] *n.* horizontal piece of wood or stone over a door or window (门或窗的)过梁,楣

longevity [lɔn'dʒeviti] *n.* long life; great duration of life; length or duration of life 长命,寿命(1)We wish you both health and longevity. 我们祝愿您俩健康长寿。(2)He prides himself on the longevity of the company. 他为公司悠久的历史而感到骄傲。

Longevity Hill 万寿山
Liang Sicheng 梁思成(1901-1972),中国著名建筑大师
Liang Qichao 梁启超(1873-1929),中国晚清著名思想家,戊戌维新运动领袖之一
Los Angeles [lɔs'ændʒələs]洛杉矶,位于美国加州西南部,美国新泽西州中部一个享有自治权的城市
Luding Bridge suspension bridge 泸定桥(索吊桥)
locate at 位于,座落在……(附近)
look like 像……似的 What do photographs look like?照片看起来是什么样子的?
lie in 在于 The solution lies in the improvement of the economic environment. 解决办法在于改善经济环境。

M

magnificent [mæg'nifisnt] *adj.* splendid or impressive in appearance; superb or very fine; (esp. of ideas) noble or elevated; great or exalted in rank or action 宏伟的,堂皇的,庄严的,华丽的(衣服,装饰等),豪华的,健壮的,优美的(体格)(1)The Taj Mahal is a magnificent building. 泰姬陵是一座宏伟的建筑。(2)She looked magnificent in her wedding dress. 她穿着婚纱,看上去漂亮极了。(3)You've all done a mangnificent job. 你们活儿干得都很出色。

mainstream ['meinstri:m] *n.* the most usual ideas or methods, or the people who have these ideas or methods 主流,主要倾向(1)His radical views place him outside the mainstream of American politics. 他的激进观点使他脱离了美国政治的主流。(2)He was never part of the literary mainstream as a writer. 作为一位作家,他从来都不属于文学的主流。

major ['meidʒə] *adj.* greater than others in importance or rank; great in number, size, or extent 1. 主要的,重要的,大的(1)a major road 一条大马路(2)major international companies 大跨国公司(3)to play a major role in sth 在某事中起重要作用(4)We have encountered major problems. 我们遇到了大问题。(5)There were calls for major changes tothe welfare system. 有人要求对福利制度进行重大改革。2.严重 Never mind, it's not major. 别担心,这不严重。

marble ['mɑ:bl] *n.* a hard crystalline metamorphic rock that takes a high polish; used for sculpture and as building material 1.大理石(1)a slab / block of marble 一块大理石板(2)a marble floor / sculpture 大理石地板 / 雕刻 2.(玻璃)弹子 3.弹子游戏 Three boys were playing marbles. 三个男孩儿在玩弹子游戏。4.理智 He's losing his marbles. 他失去理智了。

massive ['mæsiv] *adj.* imposing in size or bulk or solidity 巨大的, 结实的 a massive rock 一块巨大的岩石 the massive walls of the castle 厚实坚固的城堡围墙 The explosion made a massive hole in the ground. 爆炸在地面留下了一个巨大的坑。

masterpiece ['mɑ:stəpi:s] *n.* an outstanding work, achievement, or performance (个人或团体的)最杰出的作品，杰作, 名作(1)Thenuseum houses several of his Cubist masterpieces. 博物馆收藏了他的几件立体派杰作。(2)Her work is a masterpiece of simplicity. 她的作品是朴实的典范。

mandala ['mændələ] *n.* any of various geometric designs (usually circular) symbolizing the universe; used chiefly in Hinduism and Buddhism as an aid to meditation【宗教】曼荼罗，坛场(佛教和印度教修法地方的圆形或方形标记。一些东方国家把佛、菩萨像画在纸帛上，亦称曼荼罗)

Maya ['mɑ:jə] *n.* a member of an American Indian people of Yucatan and Belize and Guatemala who had a culture (which reached its peak between AD 300 and 900) characterized by outstanding architecture and pottery and astronomy 马雅人，马雅语 *adj.* 马雅人的,马雅语的

mechanism ['mekənizəm] *n.* the technical aspects of doing something 机制, 技巧, 原理, 途径(1)An airplane engine is a complex mechanism. 飞机引擎是种复杂的机械装置。(2)You operate the mechanism by winding this handle. 操纵这台机器要转动这个把手。(3)The mechanism of local government is far from perfect. 地方政府的结构还很不完善。(4)Meteorologists believe this pressure jump is the mechanism responsible for storms and tornadoes. 气象学家认为这一气压的突然升高是暴风雨和飓风形成的物理机制。(5)There's no mechanism for changing the decision. 没有办法改变这一决定。(6)The three sisters were all well-versed in the mechanism of novel writing. 三姐妹都熟谙写小说的技巧。

medieval [medi'i:vəl] *adj.* medieval of, relating to, or in the style of the Middle Ages; old-fashioned 中古的 connected with the Middle Ages (about AD 1000 to AD 1450)中世纪的(约公元1000到1450年)(1)The personification of evil as a devil is a feature of medieval painting. 用魔鬼象征罪恶是中世纪绘画的特色。(2)The gallery is a treasure trove of medieval art. 这个画廊是中世纪艺术的宝库。(3)This church is a classic example of medieval architecture. 这座教堂是中世纪建筑风格的典型实例。(4)The conditions were positively medieval, ie very primitive. 条件十分简陋。

miracle ['mirəkl] *n.* one that excites admiring awe; an event that appears inexplicable by the laws of nature 奇迹, 奇事(1)an economic miracle 经济方面的奇迹(2)It's a miracle (that) nobody was killed in the crash. 撞车事故中竟然没有一人丧生，这真是奇迹。(3)It would take a miracle to make this business profitable. 让这个公司赢利简直是天方夜谭。(4)a miracle cure / drug 有奇效的疗法 / 灵丹妙药

monastery ['mɔnəs,teri] *n.* a building in which monks (=members of a male religious community)live together 隐修院, 修道院, 寺院 The only regular visitors to the monastery in winter are parties of skiers who go there at Christmas and Easter. 冬季去修道院的常客只是几批滑雪者,他们通常在圣旦节和复活节去那里。

monumental [,mɔnju'mentl] *adj.* relating or belonging to or serving as a monument;of outstanding significanc 纪念的, 极为庞大的 1.重要的, 意义深远的, 不朽的(1)a monumental achiecement 重要的成就(2)Gibbon's monumental work 'The Rise and Fall of the Roman Empire' 吉本的不朽著作《罗马帝国盛衰史》2.非常大(或好、坏、蠢等)的(1)a book of monumental significance 一本意义非凡的书(2)We have a monumental task ahead of us. 极其非凡的工作正在等着我们。(3)It seems like an act of monumental folly. 这似乎是一

种非常愚蠢的行为。3. a monumental inscription / tomb 墓志铭/碑文

monuments ['mɔnjument] *n.* a notable building or site, esp. one preserved as public property 纪念碑

mummified *adj.* 木乃伊化的

mummify ['mʌʌmifai] *v.* remove the organs and dry out (a dead body) in order to preserve it 将(尸体)制成木乃伊,用香料殓藏(尸体) dry up and shrivel due to complete loss of moisture 使干枯

mural ['mjuərəl] *n.* a large painting or picture on a wall 壁画 *adj.* of or relating to a wall 墙壁似的,壁形的

Mao Zedong 毛泽东

Meiji era 明治时代

Memorial Arch 牌坊

make contribution to... 为……做出贡献 China will continue to make contribution to promote the world peace. 中国将会继续为促进世界和平而做出贡献。

more... than 与其说……,倒不如说……,不是……,而是……

N

naturalistic [,nætrrə'listik; ,nætrərə'listik] *adj.* a naturalistic style 自然的,自然主义的

nave [neiv] *n.* central space in a church, extending from the narthex to the chancel and often flanked by aisles, the long central part of a church【建】(教堂的)正殿

nourish ['nʌriʃ] *v.* to support or encourage (an idea, felling, etc) 滋养,养育 (1)We need good food to nourish the starving infants. 我们需要好食品滋养这些挨饿的婴儿。(2)Milk is all we need to nourish a small baby. 我们供给婴儿营养只需喂奶就够了。(3)Most plants are nourished by water drawn up through their roots. 多数植物是靠着根吸收水分来维持生命的。(4)She had nourished the dream of becoming a movie star. 她曾怀有成为电影明星的梦想。(5)I could find nothing to nourish my suspicion. 我找不到任何怀疑的根据。

Nijo Castle 二条城

New Urbanism【建】新都市主义 (20世纪80年代兴起的城市规划运动)

New Canaan, Connecticut [nju'kenən , kə'netikət] 康涅狄格州新迦南市,美国地名

O

OPV organic photovoltaic 有机光电伏电池

orient ['ɔ:riənt; 'əu-; 'ɔ:rient] *vt.* determine one's position with reference to another point 使朝向,以……为方向,以……为目的【建】使(尤指教堂)朝东建造 (1)Our students are oriented towards science subjects. 我们教的学生都适合学理科。(2)We run a commercially oriented operation. 我们经营一个商业性的企业。(3) profit-oriented organizations 以赢利为目的的机构 (4)Neither of them is politically oriented. 他们两人都无意涉足政治。(5)policies oriented to the needs of working mothers 针对职业母亲的需要而制定的政策

ornament ['ɔ:nəmənt] *n.* something used to beautify 1.装饰品 a china / glass ornament 瓷器 / 玻璃装饰品 2.首饰,饰物 3.装饰,摆设,点缀 The clock is simply for ornament; it doesn't work any more. 这架时钟纯属摆设,它再也不走了。*v.* Make more attractive by adding ornament, colour, etc 装饰

ornamental [,ɔ:nə'mentl] *adj.* of value as an ornament; decorative 作装饰用的,过分装饰的 (1)Modern pottery is usually ornamental. 现代陶器通常用作装饰品。(2)the art or act of decorating a text, a page, or an initial letter with ornamental designs, miniatures, or lettering 彩饰,图案花饰用装饰图案、小型画或字母

对书籍、书页或开头字母进行装饰的艺术或行为 (3)Ornamental copper pans hung on the wall.墙上挂着装饰性的铜盘。

outgrowth ['aut,grəuθ] *n.* a development,result,or consequence 自然的发展,结果(1)An outgrowth of new shoots on a branch.枝条上分出的一个新枝(2)Inflation is an outgrowth of war.通货膨胀是战争带来的后果。

overlap ['əuvə'læp] *v.* (of two things) to extend or lie partly over (each other); to cover and extend beyond (something); to coincide partly in time, subject, etc 与……交搭,叠盖住,[与……]部分相一致[巧合]

overpower [,əuvə'pauə] *v.* overcome by superior force;overcome,as with emotions or perceptual stimuli 击败(1)The robber was overpowered by the cop. 抢劫犯被警察制伏。(2)The smell of decaying meat overpowered Crompton. 腐肉的臭味使克朗普顿受不了。

on the basis of 以……为基础 Our trade is conducted on the basis of equality. 我们是在平等的基础上进行贸易。

P

pagoda [pə'gəudə] *n.* an Asian temple; usually a pyramidal tower with an upward curving roof 宝塔

parade [pə'reid] *n.* an ordered,esp.ceremonial march 游行,列队行进盛况(1)A parade was held on New Year's Day. 元旦那天举行了游行。(2)The general inspected the parade. 将军检阅了阅兵式。(3)A genuine scholar does not make a parade of his knowledge. 真正的学者不会夸耀他的知识。(4)He is always parading his knowledge. 他总是夸耀自己的知识。(5)She paraded up and down in her new hat. 她戴着一顶新帽子在人前走来走去。(6)The Olympic Games begin with a parade of all the competing nations. 奥运会以参赛各国的列队行进开始。

parallel [`pærəl] *adj.* being an equal distance apart everywhere 1.平行的(1)parallel lines 平行线(2)The road and the canal are parallel to each other.道路与运河平行。2.相似特征 There are interesting parallels between the 1960s and the late 1990s. 20世纪60年代和90年代后期存在着有趣的相似之处。3.(地球或地图的)纬线,纬圈 the 49th parallel 第49纬度线 4.in parallel (with sth / sb) (与……)同时(1)The new degree and the existing certificate courses would run in parallel. 新的学位课程和现有的证书课程将同时开设。(2)Ann wanted to pursue her own career in parallel with her husband's. 安想跟丈夫同时追求各自的事业。

parapet ['pærəpit] *n.* low protective wall along the edge of a balcony, bridge, roof, etc 矮护墙,女儿墙,(in war) protective bank of earth, stones, etc along the front edge of a trench 胸墙 He was not prepared to put his head above the parpet and say what he really thought. 他不想冒然说出自己的真实想法。

particularly [pə'tikjuləli] *adv.* in particular; specifically 特别地,格外,尤其(1)am particularly fond of her. 我特别喜欢她。(2)His behavior is not particularly adult. 他的举止行为还不太成熟。(3)The instrumentation is particularly fine. 这首器乐曲编得特别细致。(4)The house itself is not particularly to my mind, but I like its environment. 这房子本身并不特别合我的心意,但我喜欢它周围的环境。

pavilion [pə'viljən] *n.* a summerhouse or other decorative shelter 大帐蓬,亭,阁 1.(公共活动或展览用的)临时建筑物 the US pavilion at the Trade Fair 交易会上的美国展览馆 2.(运动场旁设立的)运动员席,看台 a cricked pavilion 板球队员更衣室 3.大型文体馆 the Pauley Pavilion, home of the university's basketball team 普莱文体中心,这所大学的篮球队之家 4.(公园中的)亭,阁,(音乐会、舞会的)华美建筑 his first show at the Winter Gardens Pavilion, Blackpool 他在布莱克浦冬园阁的首次演出

perimeter [pə'rimitə] *n.* the boundary line or the area immediately inside the boundary 周,周围,边缘 a line enclosing a plane areas 周长,周边 (1)Guards patrol the perimeter of the estate. 保安人员在庄园四周巡逻。

(2)a perimeter fence / track /wall 围绕四周的栅栏 / 小径 / 墙

personnel [ˌpəːsə'nel] *n.* the body of persons employed by or active in an organization, business, or service 全体人员，人事部门 Army personnel are often forbidden to fraternize with the civilian population. 军职人员常被禁止与平民百姓友好往来。Personnel is/are organizing the training of the new members of staff. 人事部门正在组织新雇员的培训。

perpendicular [ˌpəːpən'dikjulə] *adj.*【建】垂直式的

philosophy [fi'lɔsəfi] *n.* a system of philosophical inquiry or demonstration; The science comprising logic, ethics, aesthetics, metaphysics, and epistemology 1.哲学(1)moral philosophy 伦理学(2)the philosophy of science 科学原理(3)a professor of philosophy 哲学教授(4)a degree in philosophy 哲学学位 2.哲学体系，思想体系(1)the philosophy of Jung 荣格的哲学体系 (2)the development of different philosophies 不同思想体系的发展 3.人生哲学，生活的信条（或态度）Her philosophy of life is to take every opportunity that presents itself. 她的处世态度是，不放过任何呈现眼前的机会。

photovoltaic [fəutəuvɔl'teiik] *adj.* capable of producing a voltage when exposed to radiant energy, esp. light 光电(池)的，光致电压的，*n.* 光电伏打

pier [piə(r)] *n.* (1)a long structure built in the sea and joined to the land at one end(常设有娱乐场所的)突堤 (2)a long low structure built in a lake, river or the sea and joined to the land at one end(突入湖、河、海中的)码头 (3)a large strong piece of wood 柱子，墙墩，桥墩 (4)He walked along the wooden pier and climbed down into the boat. 他沿着木码头走过去并爬上了船。

pillar ['pilə] *n.* upright column of stone, wood, metal, etc used as a support or an ornament, a monument, etc 柱子 thing in the shape of this 柱形物

pioneer [ˌpaiə'niə] *n.* a colonist, explorer, or settler of a new land, region, etc.(as modifier); an innovator or developer of something new 先驱者，开拓者，拓荒者 1.pioneer (in / of sth) 先驱(1)a pioneer in the field of microsurgery 显微外科领域的创始人(2)a computer pioneer 计算机方面的先驱(3)a pioneer aviator 飞行员的先驱(4)a pioneer design 开创性设计 2. 拓荒者 the pioneer spirit 拓荒者的精神

portable ['pɔːtəbl] *adj.* carried or moved with ease 便于携带的，手提式的，可移动的(1)a portable TV 手提电视机(2)a portable loan / pension 可转移贷款 / 养老金(3)a portable software 可移动软件

portrait ['pɔːtrit] *n.* a painting of a person's face ; a picture of a person's appearance and character 肖像，半身画像(1)He had his portrait painted in uniform. 他让人画了一幅身着制服的画像。(2)a full-length portrait 全身画像(3)a portrait painter 肖像画家(4)详细的描述 a portrait of life at the French court 对法国宫廷生活的详细描述

pose [pəuz] *v.* to exist in a way that may cause a problem, danger, difficulty etc 装模作样，难题 to sit or stand in a particular position in order to be photographed or painted 摆姿势，冒充 (1)The family posed for photographs outside the house. 全家人在屋子外面摆好姿势准备拍照。(2)He posed as an artist. 他冒充艺术家。(3)Stop posing and tell us what you really think. 别装蒜啦，告诉我们你的真实想法吧。(4)Heavy traffic poses a problem in many old towns. 交通拥挤是许多旧城镇的难题。(5)His concern for the poor is only a pose. 他对穷人的关心只不过是做做样子罢了。

potent ['pəutənt] *adj.* possessing inner or physical strength; powerful; having great control or authority 1.有强效的，有力的，烈性的，影响身心的(1)a potent drug 猛药(2)a very potent alcoholic brew 烈性酒精饮料(3)a potent argument / reason 强有力的论据 / 理由 2.强大的，强有力的 a potent weapon / force 强大的武器 / 力量

precisely [pri'saisli] *adv.* in a precise manner 精确地(1)I'll tell you precisely how to do it. 我将确切地告诉你如何办理此事。(2)That answer is precisely to our need. 那恰好符合我们的需要。

predominant [pri'dɔminənt] *adj.* having superiority in power, influence, etc, over others 1.显著的,盛行的(1)a predominant feature / factor 显著特征/因素(2)Yellow is the predominant colour this spring in the fashion world. 黄色是今春时装界的流行颜色。2.占优势的,主导的(1)a predominant culture 主流文化(2)a way of thinking that is predominant in modern social life 现代社会生活中的主流思维方式

preserve [pri'zə:v] *v.* to maintain in safety from injury, peril, or harm; protect; to keep in perfect or unaltered condition; maintain unchanged 1.维护(1)He was anxious to preserve his reputation. 他急于维护自己的名声。(2)Efforts to preserve the peace have failed. 维护和平的努力失败了。2.维持原状 a perfectly preserved 14 th-century house 一座保存完好的14世纪宅第 3.保鲜(1)olives preserved in brine 盐水橄榄(2)Wax polish preserves wood and leather. 给木材的皮革打蜡起保护作用 4.使继续存活

profound [prə'faund] *adj.* [usu attrib 通常作定语] (fml 文) deep, intense or far-reaching; very great, having or showing great knowledge or insight (into a subject) 1.巨大的,深切的,深远的(1)profound changes in the earth's climate 地球气候的巨大变化(2)My father's death had a profound effort on us all. 父亲的去世深深地影响了我们全家。2.知识渊博的,深邃的(1)progound thought / understanding / insights 深邃的思想/理解/见解(2)a profound book / drama 深奥的书/戏曲

profoundly [prə'faundli] *adv.* to a great depth psychologically 1.极大地,深刻地(1)a profoundly disturbing programme 极度扰民的计划(2)We are profoundly affected by what happens to us in childhood. 童年发生的事深深地影响着我们。2.严重地,完全地,彻底地 profoundly deaf 完全失聪

project ['prɔdʒekt] *n.* a plan or proposal; a scheme; an undertaking requiring concerted effort 1.项目,方案,工程(1)a research project 研究计划(2)a building project 建筑工程(3)to set up a project to computerize the library system 开展一个图书馆系统电脑化的项目 2.(大、中学学生的)专题研究(1)a history project 历史科的专题研究(2)The final term will be devoted to project work. 最后一学期的时间将全部用于专题研究。3.方案,计划 The party attempted to assemble its aims into a focussed political project. 这个党试图把订立的目标综合为一个政治方案。4.住房 Going into the projects alone is dangerous. 只身进入公房区是危险的。

project [prə'dʒekt] *v.* to thrust outward or forward; to throw forward; hurl; to send out into space; cast 1.规划,计划,拟订方案(1)The next edition of the book is projected for publication in March. 本书的下一版计划于三月发行。(2)The projected housing development will go ahead next year. 计划中的住宅建议将于明年动工。2.预测,预计,推想(1)A growth rate of 4% is projected for next year. 预计明年的增长率为4%。(2)The unemployment rate has been projected to fall. 据预测失业率将下降。3.投射,投影 Images are projected onto the retina of the eye. 影像被投射到眼睛的视网膜上。4.突出,凸出 a building with balconies projecting out over the street 阳台伸出到街上的楼房 5.project yourself 阐述,表现(1)The sought advice on how to project a more positive image of their company. 他们就如何加强树立公司的形象征询意见。(2)She projects an air of calm self-confidence. 她表现出镇定自若的神态。

pyramid ['pirəmid] *n.* a polyhedron having a polygonal base and triangular sides with a common vertex 金字塔 1.(古埃及的)金字塔 2.(几何)锥体 3.金字塔形的物体(或一堆东西)a pyramid of cans in a shop window 商店橱窗中摆成金字塔形的罐头 4.金字塔式的组织(或系统)a management pyramid 金字塔式管理系统

Panama canal [,pænə'mɑ:'kænl] 巴拿马运河

Philip Cortelyou Johnson 菲力普·约翰逊(1906-2005),美国建筑大师

Potala Palace 布达拉宫

Princess Wencheng 文成公主

Princeton University ['prinstən] 普林斯顿大学,位于美国新泽西州的普林斯顿,是英属北美洲部分成立的第四所大学。

Pritzker Prize 普利兹克奖是每年一次颁给建筑师个人的奖项,有建筑界的诺贝尔奖之称。
provide for 供给(为……作准备,规定,考虑到) We must provide for the future. 我们必须为将来做好准备。
play a (important, significant) role in 扮演……角色 Mobile phone plays an important role in everyday life. 手机在日常生活中起着重要的作用。

Q

Qing Dynasty 清朝,中国末代王朝

quarter ['kwɔ:tə] *n.* each of four equal or corresponding parts of sth 1.四分之一(1)a quarter of a mile 四分之一英里(2)The programme lasted an hour and a quarter. 这个节目历时一小时零一刻种。(3)Cut the apple into quarters. 把苹果切成四瓣。(4)The theatre was about three quarters full. 剧场大约坐了四分之三的人。2.(整点之前或之后的)15分钟,一刻钟 It's (a)quarter to four now—I'll meet you at quarter past. 现在是差一刻四点,我会在四点一刻和你碰面。3.(士兵、服务人员等的)营房,宿舍,住房(1)We were moved to more comfortable living quarters. 我们搬进了较舒适的住处。(2)married quarters 已婚军人宿舍 *v.* divide (sb/sth) into four parts 将(人[事物])分成四部分;~ sb (on sb) provide sb with lodgings 供某人住宿 1.把……切成(或分成)四部分 She peeled and quartered an apple. 她削去苹果皮,把它切成四瓣。2.给……提供食宿 Three hundred soldiers were quartered in the town. 三百士兵在那小镇扎营。

R

radiate ['reidieit] *v.* to send out rays or wave; to manifest in a glowing manner 1.(使品质或情感)显出,流露 He radiated self-confidence and optimism. 他显得自信乐观。2.(使热、光、能量)辐射,放射,发散 Heat radiates from the stove. 炉子的热向外发散。3.(直线等)自中心辐射出(1)Five roads radiate from the square. 五条道路由广场向四处延伸。(2)The pain started in my stomach and radiated all over my body. 我起初只是肚子疼,后来是全身疼。

rafter ['rɑ:ftə] *n.* any one of a set of sloping beams that form the framework of a roof 椽

ram [ræm] *v.* ~ (against/into) sth crash against sth; strike or push sth with great force 1.和……相撞,撞击 Two passengers were injured when their taxi was rammed from behind by a bus. 公共汽车从后面撞来,出租车上的两位乘客受了伤。2.塞进,挤进(1)She rammed the key into the lock. 她将钥匙塞进锁眼。(2)The spending cuts had been rammed through Congress. 削减开支一事在国会强行通过。3.ram sth 强调(想法、论点等)以使人接受,反复灌输 4.ram into sth / ram sth into sth 猛烈撞击,使猛烈撞上另一物 He rammed his truck into the back of the one in front. 他把卡车莽地撞到前一辆卡车的车尾上。

ramp [ræmp] *n.* an inclined surface connecting two levels 1.斜坡,坡道 Ramps should be provided for wheelchair users. 应该给轮椅使用者提供坡道。2.高速公路的出口坡道 3.(装车或上下飞机的)活动梯,活动坡道 a loading ramp 装卸用的活动坡道

rampart ['ræmpɑ:t] *n.*(esp. pl.尤作复数) defensive wall. round a fort, etc consisting of a wide bank of earth with a path for walking along the top (城堡等周围宽阔的)防御土墙(esp. sing 尤作单数)

rationalist ['ræʃənəlist] *n.* someone who emphasizes observable facts and excludes metaphysical speculation about origins or ultimate causes 理性主义者,唯理主义者

rayonnant ['reiənənt]【建】(窗格等)辐射式的,光芒四射的,明亮照耀的

react [ri'ækt] *v.* show a response or a reaction to something; undergo a chemical reaction; react with another substance under certain conditions 1.react (to sth)(by doing sth) 起反应,(对……)作出反应 (1)Local residents have reacted angrily to the news. 当地居民对这一消息表示愤怒。(2)I mudged her but she didn't react. 我用胳膊肘捅了她一下,可她没有反应。 2.(对食物等)有不良反应,过敏 People can react badly to certain food additives. 人们对食品添加剂会严重过敏。3.react (with sth) / react (together) 起化学反应 Iron reacts with water and air to produce rust. 铁和水及空气发生反应产生铁锈。4.react against sb / sth 反对 He reacted strongly against the artistic conventions of his time. 他强烈反对当时的艺术俗套。

rectangular [rek'tæŋgjulə] *adj.* having four right angles; "a rectangular figure twice as long as it is wide" 矩形的,长方形的,直角的,有直角的(1)a small rectangular room 一个长方形的小房间(2)A piece of furniture for reclining and sleeping, typically consisting of a flat, rectangular frame and a mattress resting on springs. 床,一件为躺靠和睡觉用的家具,典型的由一个平的、矩形的柜架和放在弹簧上的垫子组成。(3)A rolling mill is used to form the material into various form such as round rod, square or rectangular bar. 人们使用轧机把材料加工成各种形状,如圆棒材、方棒材和矩形棒材。

readily ['redili] *adv.* without much difficulty; in a punctual manner 1.快捷的,轻而易举的,便利地 All ingredients are readily available from your local store. 所有的原料都可以方便地从你当地的商店买到。2.欣然地,乐意地 Most people readily accept the need for laws. 大多数人都毫不迟疑地认为法律是必要的。

recline [ri'klain] *v.* to rest or cause to rest in a leaning position 1.recline (against / in / on sth)斜倚,斜躺,向后倚靠(1)She was reclinning on a sofa. 她斜倚靠在沙发上。(2)a reclining figure 半躺着的人像 2.(使座椅靠背)向后倾 a reclining chair / seat 躺椅,躺式座椅

reflect [ri'flekt] *vt.* (of a mirror, etc) to form an image of (something) by reflection 1.reflect sb /sth (in sth) 反映,映出(影像)(1)His face was reflected in the mirror. 他的脸映照在镜子里。(2)She could see herself reflected in his eyes. 她在他的眼中看到了自己的样子。2.反射(声、光、热等)(1)The windows reflected the bright afternoon sunlight. 窗户反射着午后明媚的阳光。(2)When the sun's rays hit the earth, a lot of the heat is reflected back into space. 太阳光线照射到地球时,大量的热被反射回太空。3.显示,表明,表达(事物的自然属性或人们的态度、情感等)(1)Our newspaper aims to reflect the views of the localcommunity. 我们的报纸旨在表达当地人民的心声。(2)His music reflects his interest in African culture. 他的音乐反映了他对非洲文化的兴趣。4.reflect (on / upon sth) 认真思考,沉思(1)She was left to reflect on the implications of her decision. 由她负责考虑她这个决定会牵扯哪些问题。(2)On the way home he reflected that the interview had gone well. 回家的路上,他琢磨着这次面试非常顺利。5.reflect well, badly, etc. on sb / sth 使给人以好的(坏的或其他)印象 This incident reflects badly on everyone involved. 这一事件给所有相关人士都造成了恶劣影响。

regional ['ri:dʒənəl] *adj.* of, characteristic of, or limited to a region 地区性的,地域性的(1)Most regional committees meet four times a year. 大部分地区委员会每年开四次会。(2)More power is to be devolved to regional government. 要将更多的权力交给地方政府。

relic ['relik] *n.* something that has survived the passage of time, esp. an object or a custom whose original culture has disappeared; something cherished for its age or historic interest 1.遗物,遗迹,遗风,遗俗 The building stands as the last remaining relic of the town's cotton industry. 这座建筑物是小镇棉纺织业仅存的遗迹。2. holy / sacred relics 圣人遗物

relief [ri'li:f] *n.* sculpture consisting of shapes carved on a surface so as to stand out from the surrounding background【建筑学、雕塑】浮雕,浮雕品 The bronze doors are covered with sculpted reliefs. 青铜门上覆有浮雕。1.(不快过后的)宽慰,轻松,解脱(1)a sense of relief 解脱感(2)We all breathed a sight of relief when he left. 他走了以后,我们大家都如释重负地松了口气。2.relief (from / of sth) (焦虑、痛苦等的)减轻,消

除,缓和(1)modern methods of pain relief 消除疼痛的新办法(2)the relief of misery / poverty / suffering 不幸 / 穷困 / 痛楚的消除 3.(给灾区或交战地区人民提供的)救济,救援物品(1)famine relief 饥荒救济物资(2)a relief agency / organization / worker 救助机构 / 组织 / 工作者

reminiscent [ˌremi'nisnt] *adj.* stimulating memories (of) or comparisons (with) 使人联想起 (1)~ of sb /sth reminding one of or suggesting sb/sth 使人回想或联想起某人[某事物]: His style is reminiscent of Picasso's. 他的艺术风格很像毕加索的(2)The atmosphere was reminiscent of spy movies. 那气氛使人联想起间谍电影。

repetitive [ri'petitiv] *adj.* repetitious characterized by or given to unnessary repetition 重复的,反复性的(1)Has no relish for repetitive work. 不喜欢重复性的工作(2)Popular dance music characterized by strong repetitive bass rhythms. 迪斯科音乐特点是强烈重复基本旋律的一种流行舞蹈音乐。

representative [repri'zentətiv] *n.* a person or thing that represents another or others; a typical example 代表,典型代表(1)a representative of the UN 联合国代表(2)Our elected representatives in government 我们选举产生的政府代表(3)a student /union representative 学生 / 工会代表 She's our sales representative in France. 她是我们公司驻法国的销售代表。

residential [ˌrezi'denʃəl] *adj.* relating to, or having residence; suitable for, or limited to residences 1.适合居住的,住宅的 a quiet residential area 安静的住宅区 2.需要在某地居住的,提供住宿的(1)a residential language course 需要住校的语言课程(2)a residential home for the elderly 老人院(3)residential care for children 为儿童福利院提供食宿服务

resonate ['rezəneit] *v.* sound with resonance;be received or understood 1.(嗓音、乐器等)产生共鸣,回荡 Her voice resonated through the theatre. 她的声音在剧院里回荡。2.resonate (with sth)(使)回响,起回声(1)a resonating chamber 产生回音的房间(2)The room resonated with the chatter of 100 people. 屋里会合着100人唧唧喳喳的声音。3.resonate (with sb / sth) 引起共鸣 These issues resonated with the voters. 这些问题引起了投票者的共鸣。4.resonate with sth 充满 She makes a simple story resonate with complex themes and emotions. 她使一部情节简单的小说充满了复杂的主题和情感。

respective [ris'pektiv] *adj.* belonging or relating separately to each of several people or things 各自的, 分别的(1)They are each recognized specialists in their respective fields. 他们在各自的领域都被视为专家。(2)the respective roles of men and women in society 男女在社会中各自的作用

resplendent [ris'plendənt] *adj.* having a brilliant or splendid appearance 辉煌的,灿烂的,华丽的 He glimpsed Sonia, resplendent in a red silk dress. 他瞥了一眼穿着红丝绸礼服、显得光彩照人的索尼亚。

restore [ris'tɔ:] *v.* to bring back into existence or use; reestablish; to bring back to an original condition; to put (someone) back in a former position 1.restore sth (to sb) 恢复(某种情况或感受)The measure are intended to restore public confidence in the economy. 这些举措旨在恢复公众对经济的信心。2.restore sb / sth to sth 使复原使复职 He is now fully restored to health. 他现在完全恢复了健康。3.restore sth (to sth) 修复,修整,使复原 Her job is restoring old paintings. 她的工作是修复旧画。

restriction [ris'trikʃən] *n.* a principle that limits the extent of something 1.限制规定,限制法规(1)import / speed / travel restriction 进口 / 速度 / 旅行限制(2)to imposed / place a restriction on sth 对某事实行限制(3)The government has agreed to lift restrictions on press freedom. 政府已经同意撤消对新闻自由的限制。2.宽松的运动服 3.制约因素 the restrictions of a prison 监狱的种种约束

reveal [ri'vi:l] *v.* make visible; make known to the public information that was previously known only to a few people or that was meant to be kept a secret 1.reveal sth (to sb)揭示,显示,透露(1)to reveal a secret 泄露一条秘密(2)Details of the murder were revealed by the local paper. 地方报纸披露了谋杀的细节。2.显出,露

出,展示(1)He laughed, revealing a line of white teeth. 他笑起来,露出一排洁白的牙齿。(2)The door opened to reveal a cosy little room. 房门打开,一间温暖舒适的小屋展现在眼前。(3)She crouched in the dark, too frightened to reveal herself. 她蜷缩在黑暗中,吓得不敢露面。

revival [ri'vaivəl] *n.* bringing again into activit and prominence 1.(状况或力量的)进步,振兴,复苏(1)the revival of trade 贸易振兴(2)an economic revival 经济复苏(3)a revival of interest in folk music 对民间音乐的兴趣的恢复 2.复兴,再流行(1)a religious revival 宗教的复兴(2)Jazz is enjoying a revival. 爵士音乐再度盛行。

reward [ri'wɔ:d] *n./v.* a recompense for worthy acts or retribution for wrong doing 1. reward (for sth /for doing sth) 奖励,回报,报酬(1)a cash / financial reward 现金 / 财政奖励(2)a reward for good behaviour 优秀行为奖(3)The company is now reaping the rewards of their investments. 公司正在收获他们的投资回报。(4)You deserve a reward for being so helpful. 你帮了这么大的忙,理应受到奖励。(5)Winning the match was just reward for the effort the team had made. 赢得比赛的胜利是全队付出努力应得的回报。2.悬赏金 A £100 reward has been offered for the return of the necklace. 已悬赏100英镑找寻项链。

rigid ['ridʒid] *adj.* not flexible or pliant; stiff; not moving; fixed; marked by a lack of flexibility; rigorous and exacting 1.(规则、方法等)死板的,僵硬(1)The curriculum was too narrow and too rigid. 课程设置过于狭窄和死板。(2)His rigid adherence to the rules made him unpopular. 他对规则的刻板坚持使得他不受欢迎 2.固执的,一成不变的 rigid attitudes 固执的态度 3.(物体或物质)坚硬的,僵直的(1)a rigid support for the tent 帐篷的坚硬的支柱(2)She sat up right, her body rigid with fear. 她直挺挺地坐着,吓得浑身发僵。

rival ['raivəl] *n.* a person, a team,etc, that competes with another for the same object or in the same field 对手,竞争者(1)They are rivals for the same position. 他们是争夺同一个职位的敌手。(2)His collection of stamps has few rivals in the world. 他搜集的邮票几乎举世无双。(3)The party leader has been supplanted by his rival. 那位政党领导已被他的对手取而代之了。

Romanesque [rəumə'nesk]*adj.*【建】罗马式建筑,罗马式绘画[雕刻]

ruin ['ru:in] *n./v.* the parts of a building that remain after it has been destroyed or severely damaged 残垣断壁 to damage something so badly that it loses all its value,pleasure,etc 毁坏,破坏,糟蹋 to make somebody/something lose all their money,their position,etc 使破产(或失去地位等),毁灭(1)476AD saw the ruin of Roman Empire. 公元四七六年罗马帝国灭亡。(2)We visited the ruins of the temple. 我们参观了那个庙宇的遗迹。(3)The news meant the ruin of all our hopes. 这消息使我们的一切希望都破灭了。

rustic ['rʌstik] *n.* an unsophisticated country person 乡下人 *adj.* characteristic of rural life; awkwardly simple and provincial 乡村的(1)an old cottage full of rustic charm 充满了乡村魅力的旧农舍(2)用粗糙木材做成的 a rustic garden seat 花园里的粗木座椅

Richard Meier 理查德.迈耶(1934-),美国建筑大师。

range from ...to 范围从……到…… Their ages range from 25 to 50. 他们的年龄在25岁到50岁之间。

rather than 而不是 Rational rather than emotional 理性的而不是感性的

react against 反抗 He reacted against his father's influence by becoming a priest. 他为了反抗他父亲的影响当了传教士。

refer to 提到 Don't refer to this matter again, please. 请不要再提这件事。

S

sandstone ['sændstəun] *n.* a sedimentary rock consisting of sand consolidated with some cement

savor ['seivə] *n.* the taste experience when a savoury condiment is taken into the mouth 味道, 气味, 滋味 For Emma, life had lost its savor. 对埃玛来说, 生活已失去乐趣。 *v.* derive or receive pleasure from; get enjoyment from; take pleasure in 品尝……滋味

secular ['sekjulə] *adj.* not concerned with or related to religion 现世的, 尘世的, 世俗的, 非宗教的

secure [si'kjuə] *v.* get by special effort 保护, 招致, 弄到 (1)He secured all the windows before he left. 他走之前关紧了所有的窗户。(2)The little boy felt secure near his parents. 那小男孩在父母身边感到安心。(3)Some measures are needed to secure the farmland against shifting sand. 需要采取措施使农田免遭流沙的侵袭。(4)He secured a position in a store at US $400 a week. 他在一家商店获得了一份每周四百美元的工作。(5)His words secured his father's displeasure. 他的话招致他父亲的不悦。

scalable [skeiləbl] *adj.* 可伸缩的, 可攀登的, 可升级的

screen [skri:n] *n.* something that keeps things out or hinders sight 屏幕 (1)He hid behind the screen and overheard their conversation. 他躲在帘子后面, 偷听到他们的谈话。(2)We have screens on our windows to keep out the flies. 我们窗户上有纱窗以挡苍蝇。(3)A screen of trees hides the villas from the beach. 这些别墅有树木遮掩, 从海滩上是看不见的。

sculpt [skʌlpt] *v.* to practice sculpture 雕刻 (1)I would have burned the pink candle sculpted like a rose before it melted in storage. 我会点燃那支雕成玫瑰状的蜡烛, 而不让它在尘封中熔化。(2)Sculpting brings her pleasure and extra income. 雕塑是她兴趣所在, 并带来额外收入。

sculpture ['skʌlptʃə] *n.* the art of making figures or designs in relief or the round by carving wood, moulding plaster, etc, or casting metals, etc; works or a work made in this way 1.雕刻, 雕塑品 (1)a marble sculpture of Venus 维纳斯的大理石雕像 (2)He collects modern sculpture. 他收藏现代雕塑。2.雕刻术 the techniques of sculpture in stone 石雕技艺

shade [ʃeid] *n.* cover or shelter provided by interception by an object of the sun or its rays 遮光物, 罩, 幕[pl.][美俚]太阳眼镜 (1)I saw him sitting in the shade of a tree. 我看见他坐在树荫下。(2)There are no trees or bushes to give shade. 没有树木或灌木丛可以遮荫。

shrine [ʃrain] *n.* a place of worship hallowed by association with some sacred thing or person 圣地, 神龛, 庙, 圣祠 (1)a shrine to the Virgin Mary 敬奉圣母玛利亚的朝圣地 (2)to visit the shrine of Mecca 前往圣地麦加朝拜 (3)Wimbledon is a shrine for all lovers of tennis. 温布尔登是所有网球爱好者的圣地。*v.* enclose in a shrine 将……置于神龛内

simplicity [sim'plisiti] *n.* the quality or condition of being simple 简朴, 朴素, 简单 (1)The advantage of the idea was its simplicity. 这个主意的优点就在于它简单明了。(2)She dressed with elegant simplicity. 她的穿着朴素而雅致。(3)The beauty of the plan consists in its simplicity. 该计划的妙处在于简洁明了。(4)The monastic community generally lives a life of simplicity. 僧侣界通常过着简朴生活。(5)Simplicity is the essence of good taste. 纯朴是情趣高尚的主要因素。

simultaneous [ˌsiməl'teinjəs] *adj.* occurring or operating at the same time 同时的 (1)There were several simultaneous attacks by the rebels. 反叛者同时发动了几起攻击。(2)Any ceasefire would be simultaneous with the withdrawal of US forces. 美军何时撤走, 停火便在何时实施。(3)simultaneous translation / interpreting 同声传译

solid ['sɔlid] *adj.* hard or firm; not in the form of a liquid or gas 固体的, 坚硬的 strong and made well 结实的, 坚

固的, 牢固的(1)When water freezes, it becomes solid. 水结冰时变成固体。(2)When water freezes it becomes solid and we call it ice. 水遇冷凝结, 称之为冰。(3)This horse has good solid muscle on him. 这匹马长得很结实。

spatial ['speiʃəl] *adj.*pertaining to or involving or having the nature of space 空间的, 存在于空间的, 受空间条件限制的

spectacle ['spektəkl] *n.*a strange or interesting object or phenomenon（引人注目的）景象, 奇观(1)The display of fireworks on New Year's Eve was a fine spectacle. 除夕燃放的烟火真是美妙的奇观。(2)The celebrations provided a magnificent spectacle. 庆祝活动呈现一派宏伟的景象。(3)The sunrise seen from high in the mountains was a tremendous spectacle. 从山上居高远望, 日出景象蔚为奇观。

spectacular [spek'tækjulə] *adj.* of the nature of a spectacle; impressive or sensational.引人注目的, 壮观的
(1) spectacular scenery / views 壮丽的风光 / 景色(2)Giggs scored a spectacular goal. 吉格斯进了一个球, 精彩极了。(3)It was a spectacular achievement on their part. 这是他们取得的一项了不起的成就。

spectrum ['spektrəm] *n.* the distribution of energy emitted by a radiant source, arranged in order of wavelengths. 光谱(1)A spectrum is formed by a ray of light passing through a prism. 一束光通过棱镜就会形成光谱。(2) Red and violet are at opposite ends of the spectrum. 红色和紫色位于光谱的两端。(3)the electromagnetic / radio / sound spectrum 电磁波谱, 射频频谱, 声谱

speculation [ˌspekju'leiʃən] *n.* a message expressing an opinion based on incomplete evidence 推测, 猜测, 推断
(1)There was widespread speculation that she was going to resign. 人们纷纷推测她将辞职。(2)His private life is the subject of much speculation. 他的私生活引起诸多猜测。(3)Today's announcement ends months of speculation about the company's future. 今天的声明使得几个月来关于公司未来的种种猜测就此烟消云散。(4)She said dismissed the newspaper reports as pure speculation. 她说报纸上的报道毫无根据, 纯属臆断。

sphinx [sfiŋks] *n.* one of a number of large stone statues with the body of a lion and the head of a man that were built by the ancient Egyptians 狮身人面像, 谜一样的人

starter ['stɑ:tə] *n.*调度员, （自动）起动机, 起[启]动器, 启动装置(1)Of the five starters in the race only three finished. 起跑时有五个,只有三个跑完全程。(2)He's a fast starter. 他做事起步很快。(3)remoter starter 遥控器

strategic ['strəti:dʒik] *adj.*[usu attrib 通常作定语] of strategy; forming part of a plan or scheme 战略（上）的, 策略（上）的 (of weapons, esp. nuclear missiles) directed against an enemy's country rather than used in a battle (指武器, 尤指核武器)战略性的 strategic missiles 战略导弹

striking [ˌstraikiŋ] *adj.* attracting attention; fine; impressive 1.引人注目的, 异乎寻常的, 显著的(1)a striking feature / example 一个引人关注的特征 / 例子(2)She bears a striking resemblance to her older sister. 她酷似她姐姐。(3)In striking contrast to their brothers, the girls were both intelligent and charming. 姑娘们既聪明伶俐, 又妩媚动人, 跟她们的兄弟形成鲜明的对照。2.妩媚动人的, 标志的, 俊秀的(1)striking good looks 姣好的面容(2)She was undoubtedly a very striking young woman. 她无疑是一个很漂亮的年轻女子。

strive [straiv] *v.* to exert much effort or energy; endeavor; to struggle or fight forcefully; contend 努力, 奋斗, 争斗争 (1)We encourage all members to strive for the highest standards. 我们鼓励所有成员为达到最高标准而努力。(2)striving against corruption 与腐败现象进行斗争(3)Newspaper editors all strive to be first with a story. 报纸编辑部都力争率先报道。

stucco ['stʌkəu] *n.* a plaster now made mostly from Portland cement and sand and lime; applied while soft to cover exterior walls or surfaces 灰泥,（涂墙壁或天花板用的）粉饰灰泥

subdivide ['sʌbdi'vaid] *v.* to divide(somthing)resulting from an earlier division 再分，细分~ (sth) (into sth) (1) The house is being subdivided into apartments. 这房子正被分成公寓套间。(2)Part of the building has been subdivided into offices. 这座大楼的一部分隔开用作办公室了。(3)This division of the chapter has several subdivisions. 该章这一节里又分成几个小节。

succession [sək'seʃən] *n.* a following of one thing after another in time; acquisition of property by descent or by will 连续,继承权,继位 (1)a succession of events / problems / visitors 接二连三的事情 / 一系列问题 / 络绎不绝的客人(2)He's been hit by a succession of injuries since he joined the team. 自入队以来他一再受伤。(3)She has won the award for the third year in succession. 这是她连续第三年获得此奖。(4)She's third in order of succession to the throne. 她在王位继承人顺位中排第三。

successive [sək'sesiv] *adj.* following another without interruption; of or involving succession 接连的,相继的,继[连]续的,逐次[步]的(1)This was their fourth successive win. 这是他们连续第四次获胜。(2)Successive governments have tried to tackle the problem. 历届政府都试图解决这个问题。(3)There has been low rainfall for two successive years. 连续两年降雨量偏低。

superb [sju:'pə:b] *adj.* of surpassing excellence; surpassingly good 宏伟的,壮丽的,华美的,极佳的,卓越的 (1) a superb player / meal / goal 一名杰出的运动员 / 一顿丰盛的饭 / 一个精彩的入球(2)The car's in superb condition. 这辆车车况极好。(3)His performance was absolutely superb. 他的表演精彩绝伦。

suppress [sə'pres] *v.* to put an end to; prohibit 镇压,抑制,查禁；使止住 (1)The rebellion was brutally suppressed. 起义遭到了残酷的镇压。(2)The police were accused of suppressing vital evidence. 警方被指隐瞒关键证据。(3)She was unable to suppress her anger. 她按捺不住怒火。

surpass [sə:'pɑ:s] *v.* to be beyond the limit, powers, or capacity of; transcend; to be or go beyond, as in degree or quality; exceed 超越,胜过(1)He hopes one day to surpass the world record. 他希望有一天能刷新世界记录。(2)Its success has surpassed all expectations. 它所取得的成功远远超出了预期。(3)Her cooking was always good, but this time she had surpassed herself. 她的厨艺向来不错,但这一次她更是胜过以往。(4) scenery of surpassing beauty 无比优美的景色

surrender [sə'rendə] *v.* ~ (oneself) (to sb) stop resisting an enemy, etc; yield; give up 停止抵抗,投降,屈服,放弃，~ sth/sb (to sb) (fml 文) give up possession of sth/sb when forced by others or by necessity; hand sth/sb over 被迫放弃对某物(某人)的控制权,交出某事物[某人](1)The rebel soldiers were forced to surrender. 叛军被迫投降。(2)The hijackers eventually surrendered themselves to the police. 劫机者最终向警方投降。

suspend [sə:s'pend] *v.* to hang so as to allow free movement; to cause to stop for a period; interrupt 1.吊,挂,中断(1)A lamp was suspended from the ceiling. 一盏吊灯悬在天花板上。(2)Her body was found suspended by a rope. 人们发现她的尸体吊在绳子上。2.暂停；使暂停发挥作用(或使用等)(1)Production has been suspended while safety checks are carried out. 在进行安全检查期间生产暂停。(2)The constitution was suspended as the fighting grew worse. 鉴于战斗趋于激烈,宪法暂停实施。3.延缓,暂缓,推迟(1)The introduction of the new system has been suspended until next year. 新制度推迟到明年再行实施。(2)to suspend judement 暂不判断

suspension [səs'penʃən] *n.* the act of suspending or the condition of being suspended 1.暂令停职(或停学、停赛等)(1)suspension from school 暂被停学(2)The two players are appealing against their suspensions. 这两名运动员请求取消对他们的停赛处罚。2.暂缓,推迟,延期 These events have led to the suspension of talks. 这些事件导致谈判延期。3.(车辆减震用的)悬架 the front / rear suspension 前 / 后悬架4.悬浮液,悬浮

symbolism ['simbəlizəm] *n.* a system of symbols and symbolic representations; an artistic movement in the late 19th century that tried to express abstract or mystical ideas through the symbolic use of images 象征,记号,

印象派绘画艺术

symmetry ['simitri] *n.* exact match in size and shape between the two halves of sth pleasingly regular way in which parts are arranged 匀称,对称

Shaanxi Loess Plateau 陕西黄土高原

Shang Dynasty 商朝

Songtsen Gampo 松赞干布

stilt houses 吊脚楼

Suez canal ['sju(:)iz] 苏伊士运河

Suzaku Road 朱雀路

scores of 大量,许多 Scores of guests had been trapped in their rooms, too terrified to move. 许多客人被困在房间里吓得走不动。

seek for 寻找,追求,探索 Young people like to seek for success in life. 年轻人喜欢探索人生的成功之途。

separate... into... 分成…… A chemist can seperate a medicine into its components. 化学家能把一种药物的各种成分分解出来。

serve as... 作为,用作 This will serve as certification. 以此为凭。

serve to 用来(足以) This success will only serve to spur her on. 这个成绩只会鼓舞她继续前进。

side by side 并排,并肩 two children walking side by side 两个并肩行走的小孩

T

Taoism 道教

tectonic [tek'tɔnik] *adj.* relating to construction or building 构造的,建筑的

temple ['templ] *n.* a building or place dedicated to the worship of a deity or deities 1.(非基督教的)庙宇,寺院,神殿,圣堂(1)the Temple of Diana at Ephesus 以弗所的狄安娜神庙(2)Buddhist / Hindu / Sikh temple 佛教/印度教/锡克教庙宇(3)to go to temple 去会堂礼拜 2.太阳穴,鬓角 She had black hair, greying at the temples. 她头发乌黑,但两鬓渐白。

territory ['teritəri] *n.* a region marked off for administrative or other purposes; an area of knowledge or interest 1.领土,版图(1)enemy / disputed / foreign territory 敌方/有争议的/外国领土(2)occupied territories 被占领的土地(3)They have refused to allow UN troops to be stationed in their territory. 他们拒不允许联合国部队驻扎在他们的国土上。2.(个人、群体、动物等占据的)领域,管区,地盘(1)Mating blackbirds will defend their territory against intruders. 乌鸫交配时会保护自己的底盘,不允许外来者侵入。(2)This type of work is uncharted territory for us. 我们从未涉足过这类工作。3.(某人负责的)地区 Our representatives cover a very large territory. 我们的代理人负责的地区很广。4.(某类)地区,(某种)地方 unexplored territory 未勘察地区

texture ['tekstʃə] *n.* the feel of a surface or a fabric; the essential quality of something 1.质地,手感(1)the soft texture of velvet 天鹅绒柔软的质地(2)She uses a variety of different colours, shapes and textures in her wall hangings. 她悬挂的帷幔不论颜色、形状和质地都多姿多彩。2.口感 The two cheeses were very different in both taste and texture. 这两种奶酪的味道和口感大不相同。3.(音乐或文学的)和谐统一感,神韵 the rich texture of the symphony 这首交响曲优美和谐的乐感

tie [tai] *v.* to fasten or secure with or as if with a cord, rope, or strap; to bring together in relationship; connect or unite 1.(用线、绳等)系,拴,绑,捆,束(1)She tied the newspapers in a bundle. 她把报纸扎成一捆。(2)He had to tie her hands together. 他不得不把她的双手绑在一起。2.将……系在……上 (1)She tied a label on to the suitcase. 她把签条系在衣箱上。(2)He tied on apron on and got down to work. 他系上围裙就开始干活。3.(在线、绳上)打结,系扣(1)to tie a ribbon / tie 系丝带 / 领带(2)Tie up your shoelaces！把你的鞋带系好！4. 使紧密结合(1)Pay increases are tied to inflation. 提高工资和通货膨胀紧密相关。(2)The house is tied to the job, so we'll have to move when I retire. 这房子是为工作提供的,所以我退休后我们就得搬家。

tile [tail] *n.* a flat thin slab of fired clay, rubber, linoleum, etc, usually square or rectangular and sometimes ornamental, used with others to cover a roof, floor, wall, etc 1. 瓷砖,地砖,小方地毯,片状材料2.瓦,瓦片3.棋子

timber ['timbə] *n.* wood used as a building material; lumber 1.(用于建筑或制作物品的)树木,林木,用材林 the timber industry 林木业2.(建筑等用的)木材,料 houses built of timber 木屋3.(建造房屋用的)大木料,栋木,大梁4.(造船用的)船骨,肋材 roof timbers 房檩

throne [θrəun] *n.* the chair of state of a monarch, bishop, etc 1.(国王、女王的)御座,宝座2.王位,王权,帝位 Queen Elizabeth came / succeeded to the throne in 1952. 伊丽莎白女王于1952年即位 / 登基。

tracery ['treisəri] *n.* a pattern of interlacing ribs, esp. as used in the upper part of a Gothic window, etc【建】窗花格,花格子,由线纹构成的装饰细工或图样

transform [træns'fɔ:m] *v.* change or alter in form, appearance, or nature 变换,改变,转化 (1)The Greggs have transformed their garage into a guest house. 格雷格一家把他们的车库改成了客房。(2)His plans were transformed overnight into reality. 他的计划迅速变为现实。(3)The magician transformed the frog into a princess. 魔术师把青蛙变成了公主。(4)Success and wealth transformed his character. 成功和财富改变了他的性格。

transport [træns'pɔ:t] *v.* to carry from one place to another; convey 传送, 运输 the act of transporting; conveyance *n.* 运输, 运输机 1.(用交通工具)运输,运送,输送(1)to transport goods / passengers / cattle 运送货物 / 旅客 / 牛(2)Most of our luggage was transported by sea. 我们的大部分行李都是海运的。2.(以自然方式)运输,输送,传播(1)The seeds are transported by the wind. 这些种子是由风传播的。(2)Blood transports oxygen around the body. 血把氧气输送到全身。3.使产生身临其境的感觉 The book transports you to another world. 这本书会把你带到另一个世界。4.(旧时)流放

trench [trentʃ] *n.* a ditch dug as a fortification having a parapet of the excavated earth 沟,沟渠,堑壕(1)life in the trenches 第一次世界大战的战壕生活(2)trench warfare 堑壕战 *v.* fortify by surrounding with trenches (挖)沟,(挖)战壕

triangle ['traiæŋgl] *n.* a three-sided polygon 三角形,三角形物体 a right-angled triangle or a right triangle 直角三角形

trickle ['trikl] *v./n.* to flow or fall in drops or in a thin stream; to move or proceed slowly or bit by bit 滴,细流,缓慢移动(1)Tears were trickling down her cheeks. 眼泪顺着她的面颊流了下来。(2)Trickle some oil over the salad. 往色拉上滴些油。

trimmer ['trimə] *n.* a device or machine, such as a lumber trimmer, that is used for trimming 调整者,装饰者,修剪器,剪切器

trustee [trʌs'ti:] *n.* a person to whom the legal title to property is entrusted to hold or use for another's benefit; a member of a board that manages the affairs and administers the funds of an institution or organization 1.(财产的)受托人 The bank will act as trustees for the estate until the child is 18. 这家银行将充当遗产的受托人,直到这孩子18岁为止。2.(慈善事业或其他机构的)受托人

tunable ['tju:nəbl] *adj.* that can be tuned 可调音的,和谐的,悦耳的

tunnel ['tʌnl] *n.* an underground passageway, esp. one for trains or cars that passes under a mountain, river, or a congested urban area 1.地下通道,地道,隧道(1)a railway / railroad tunnel 铁路隧道(2)the Channel Tunnel 英吉利海峡隧道 2.(动物的)洞穴通道

turbulent ['tə:bjulənt] *adj.* violently agitated or disturbed; tumultuous 1.动荡的,动乱的,骚动的,混乱的(1)a short and turbulent caereer in politics 短暂动荡的政治生涯(2)a turbulent part of the world 世界上动荡不安的地区 2.汹涌的,猛烈的,湍动的(1)The aircraft is designed to withstand turbulent conditions.这架飞机是为经受猛烈的气流而设计的。(2)a turbulent sea / storm 波涛汹涌的大海/狂风暴雨

tycoon [tai'ku:n] *n.* a very wealthy or powerful businessman 巨富 a business / property / media tycoon 产业大亨,房地产巨头,传媒巨富

Taj Mahal ['tɑ:dʒmə'hɑ:l] beautiful mausoleum at Agra built by the Mogul emperor Shah Jahan (completed in 649) in memory of his favorite wife 泰吉马哈尔陵(亦译泰姬陵,印度阿格拉的一座大理石陵墓,由17世纪莫卧儿帝国皇帝 Shah Jahan 为爱妃所建)

the Bank of China Tower 香港中国银行大厦,由贝聿铭建筑师事务所设计,1990年完工。

the Chinese University of Hong Kong 香港中文大学

the Expo Forums 世博会论坛

the Grand Canal 京杭大运河

the Great Depression 美国经济大萧条时期 (1929-1930s)

the Great Wall 长城

the National Gallery of Art (美国)国家美术馆,美国国家美术馆位于美国国会大厦西阶,国家大草坪北边和宾夕法尼亚大街夹角地带,是世界上建筑最精美、藏品最丰富的美术馆之一。

the National Grand Theater (中国)国家大剧院

the New York Five 纽约五人派,纽约五人组,又称"白派",是现代派原则的拥护者。代表人物有理查·迈耶、迈克·格雷夫斯、查理士·盖斯密、彼得·艾森曼及约翰·黑达克。

the Summer Palace 颐和园

the Yangtze River 扬子江

Tiananmen Square 天安门广场

Tian Xia Di Yi Guan 天下第一关(包括山海关城、东罗城城楼、靖边楼、牧营楼、临闾楼等)

take along 随身带着,随……带来 You'd better take your umbrella along. 你最好把雨伞带着。

take place 发生,举行 When will the basketball game take place? 篮球赛何时举行?

take the throne 登基为王,加冕为王

tend to 有……的倾向 Modern furniture design tends to simplicity. 现代家具设计越来越简单。

to this day 至今,迄今,直到现在

U

ultimate ['ʌltimit] *adj.* fundamental; elemental; of the greatest possible size or significance; maximum 最后的,最终的,终极的(1)Our ultimate goal / aim / objective / target 我们最终的目的;目标(2)We will accept ultimate responsibility for whatever happens. 无论出什么事情,我们愿承担全部责任。(3)the ultimate truths of philosophy and science 哲学与科学的终极原理

ultimately ['ʌltimətli] *adv.* as the end result of a succession or process 最后,最终(1)Ultimately, you'll have tomake the decision yourself. 最终你还是得自己拿主意。(2)A poor diet will ultimately lead to illness. 不

均衡的饮食终将导致疾病。

underworld ['ʌndəwə:ld] *n.* the criminal class; (religion) the world of the dead 地狱,下层社会,尘世,阴曹地府 (1)the criminal underworld 罪恶的黑社会 (2)the Glasgow underworld 格拉斯哥的黑社会。

unified ['ju:nifaid] *adj.* unify 的过去式和过去分词 operating as a unit 统一的,统一标准的

unique [ju:'ni:k] *adj.* single; sole without equal or like; unparalleled; very remarkable or unusual; leading to only one result 1.唯一的,独一无二的 Everyone's fingerprints are unique. 每个人的指纹都是独一无二的。2.独特的,罕见的(1)a unique talent 奇才 The deal will put the company in a unique position to export goods to Eastern Europe. 这项协议给予这家公司向东欧输出商品的特殊地位。3.(某人、地或事物)独具的,特有的

unparalleled [ən'pærəleld] *adj.* radically distinctive and without equal 无比的,优良无比的,空前的 (1)It was an unparalleled opportunity to develop her career. 这是她发展自己事业的绝好机会。(2)The book has enjoyed a success unparalleled in recent publishing history. 这本书在近期出版史上是空前的成功。

up-to-date ['ʌptə'deit] *adj.* reflecting the latest information or changes; in accord with the most fashionable ideas or style 1.现代的,时髦的,新式的(1)This technology is bang up to date. 这项技术是最新式的。(2)up-to-date clothes 时髦服装(3)up-to-date equipment/methods 最新的设备/方法 2.拥有（或包含）最新信息的(1)We are keeping up to date with the latest developments. 我们保持掌握最新的情况发展。(2)up-to-date information / records 最新的信息 / 记录

urbanization [,ə:bənai'zeiʃən] *n.* the social process whereby cities grow and societies become more urban 都市化

usher ['ʌʃə] *v.* show (someone) to their seats, as in theaters or auditoriums 引领,招待,陪同,迎接,预报……的来到,开辟,创始 The secretary ushered me into his office. 秘书把我领进他的办公室。

United Nations headquarters 联合国总部,设在美国纽约市曼哈顿区哈得逊河边。

UNESCO 联合国教科文组织 (United Nations Educational, Scientific and Cultural Organization 联合国教育科学及文化组织)

V

vague [veig] *adj.* not clearly understood or expressed; not precisely limited, determined, or distinguished 1.(思想上)不清楚的,含糊的,不明确的,模糊的 2.vague (about sth) 不具体的,不详细的,粗略的 (1)She's a little vague about her plans for next year. 她对明年的计划不怎么明确。(2)He was accused of being deliberately vague. 他被指责为故意含糊其辞。3.(人的行为)茫然的,糊涂的,心不在焉的(1)a vague expression / look 茫然若失的表情/神情(2)His vague manner concealed a brilliant mind. 他大智若愚。4.朦胧的 In the darkness they could see the vague outline of a church. 他们在黑暗中能看到一座教堂的朦胧轮廓。

variety [və'raiəti] *n.* a collection containing many kinds of things 1.(同一事物的)不同种类,多种式样 There is a wide variety of patterns to choose from. 有种类繁多的图案可供选择。2.变化,多样化,多变性 We want more variety in our work. 我们希望我们的工作多变点儿花样。3.variety (of sth) (植物、语言等的)变种,变体,品种 a rare variety of orchid 兰花的稀有品种 4.The variety is the spice of life. 经历丰富多彩令生活充满乐趣。

vault [vɔ:lt] *n.* 1. a room with thick walls and a strong door where money, jewels etc are kept to prevent them from being stolen or damaged 金库 2. a room where people from the same family are buried, often under the floor of a church (教堂的)地下墓室,墓穴 3. a jump over something 撑竿跳 4.a roof or ceiling that consists of several arches that are joined together, esp. in a church【建】拱顶,穹隆

vent [vent] n. 1. a hole or pipe through which gases, liquid etc can enter or escape from an enclosed space or container: (空气、气体、液体的)出口, 进口, 漏孔 air / heating vents 通气 / 热风孔 2. the small hole through which small animals, birds, fish, etc pass waste matter out of their bodies: (鸟、鱼等小动物的)肛门 3. a thin straight opening at the bottom of the back or side of a jacket or coat: 衩口, 开衩, 背衩 4. give vent to sth (fml 文) to do something violent or harmful to express feelings of anger, hatred etc: (充分)表达, (淋漓尽致地)发泄 She gave full vent to her feelings in a violent outburst. 她大发脾气以宣泄情绪。

version ['vəːʃən] n.1. a copy of something that has been changed so that it is slightly different version of: a new (1)version of the software 软件版(2)new/modern/final etc version 最新/最时尚/.最终等版本(3)the original version of the text 原版课文(4) English/German/electronic/film etc version (=presented in a different language or form) 英语/德语/电子/电影等版本 2. someone's version of an event is their description of it, when this is different from the description given by another person (个人对事件的)描述, 看法, 说法 Could Donnas' version of what happened that night be correct? 那天晚上杜勒斯讲述的事件是真的吗？ 3 a way of explaining or doing something that is typical of a particular group or period of time 学说 the Marxist version of economics 马克思主义经济学说

Victorian Era [vik'tɔːriən] ['iərə] 维多利亚时代(1837-1901)

victory ['viktəri] n.final and complete superiority in a war 胜利, 战胜, 克服(1)The basketball team had a string of victories last season. 这个篮球队在上一个赛季中赢得了一连串的胜利。(2)We win the emphatic victory.我们赢得了那场有目共睹的胜利。(3)The soldiers sang a song of victory, describing their prowess in battle. 战士们唱起了歌颂他们英勇战斗的凯歌。(4)I fervently believe in our eventual victory.我坚信我们最后一定会胜利。

vision ['viʒən] n.a mental image produced by the imagination; the faculty of sight; eyesight 梦想, 幻想, 幻觉, 视力, 视觉, 想象力(1)People wear glasses to improve their vision. 人们戴眼镜以改善视力。(2)The vision of the table loaded with food made our mouths water. 看见桌上堆满了食物使我们垂涎欲滴。(3)He is a man of great vision. 他是位有远见的人。

vital ['vaitl] adj.most important,necessary to continued existence or effectiveness; essential 极其重要的, 必不可少的, 致命的, 生死攸关的, 严重的(1)Growth and decay are vital processes. 生长和衰亡是生命过程。(2)The heart is a vital organ.心脏是维持生命必需的器官。(3)The Chinese I knew were trusting, open, and vital. 我所认识的中国人信赖别人, 坦率, 充满活力。

vogue [vəug] n. ~for sth,a fashion for sth. 尚, 流行, 时髦(1)Short hair came back into vogue.短发又开始流行起来了。(2)a new vogue for low-heeled shoes 低跟鞋新潮流.(3)Full-length coats are no longer the vogue. 长外套不再流行了。(4)His novels were in vogue ten years ago.他的小说十年前风靡一时。

W

warm up 热身, 变暖, 感到亲切, 激动

watch-towers 角楼

wisdom ['wizdəm] n. 1. good sense and judgment, based esp. on your experience of life:(1) a man of great wisdom 智者(2)question/doubt the wisdom of (doing) something: Local people are questioning the wisdom of spending so much money on a new road.当地群众对投巨资修建一条公路产生质疑。 2. knowledge gained over a long period of time through learning or experience: the collected wisdom of many centuries 几百年积累下来的知识(学问)3. (the) conventional/received/traditional etc wisdom 多数人的看法,普遍信念

The conventional wisdom is that boys mature more slowly than girls. 人们普遍认为男孩比女孩成熟得晚。 4. in somebody's (infinite) wisdom (humorous used to say that you do not understand why someone has decided to do something) (表示不理解他人的) 无知, 愚蠢 The boss, in her infinite wisdom, has decided to reorganize the whole office yet again. 老板竟愚蠢到要再次重组各管理部门。

World Exposition 世界博览会
World Heritage 世界遗产
World Heritage List 世界文化遗产名录
World War II 第二次世界大战 (1939-1947)

Appendix 4 Architectural Technical Terms 建筑工程术语

aseismic joint 防震缝
arch structure 拱结构
arch 拱
accidental situation 偶然状况
action 作用
accidental combination for action effects 作用效应偶然组合
area of section 截面面积
accidental action 偶然作用
amplitude of vibration 振幅
at an angle to grain 斜纹
accidental combination 偶然组合
assembled monolithic concrete structure 装配整体式
accidental load 偶然荷载
acceleration 加速度
allowable subsoil deformation 地基变形允许值
autoclaved sand-lime brick 蒸压灰砂砖
autoclaved flyash-lime brick 蒸压粉煤灰砖
acceptance 验收
ambient temperature 环境温度
a separate waterproof barroer 一道防水设防
border adhibiting method 空铺法
built-up member 组合构件
building engineering 建筑施工
bracing member 支管
buckling 屈服
brittle fracture 脆断
box foundation 箱形基础
basic variable 基本变量
beam; girder 梁
beam fixed at both ends 两端固定梁
bent frame 排架
bending moment 弯矩
bimoment 双弯矩
building and civil engineering structures 工程结构

building engineering 房屋建筑工程
building 房屋建筑
building decoration 建筑装饰装修
brittle failure 脆性破坏
breadth of section 截面宽度
building height 房屋高度
building ground 建筑地面
beam end 梁端有效支承长度
base course 基层
calibration 校准法
cantilever beam 悬臂梁
cable-suspended structure 悬索结构
civil engineering 土木工程
civil building; civil architecture 民用建筑
characteristic value of a property of a material 材料性能标准值
compressive capacity 受压承载能力
coefficient of effects of actions 作用效应系数
combination for shourt-term action effects 短期效应组合
combination for long-term action effects 长期效应组合
combination for action effects 作用效应组合
compressive strength 抗压强度
column 柱
component; assembly parts 部件
construction joint 施工缝
construction works 建筑物(构筑物)
connection 连接
continuous beam 连续梁
concrete structure 混凝土结构
combination value of actions 作用组合值
coefficient of friction 摩擦系数
cylinder pile foundation; cylinder caisson foundation 管柱基础

caisson foundation 沉箱基础

characteristic value of subgrade bearing capacity 地基承载力特征值

cast-in-situ concrete structure 现浇混凝土结构

characteristic value of strength 强度标准值

characteristic combination 标准组合

cantilever beam 挑梁

calculating overturning point 计算倾覆点

composite steel and concrete beam 钢筋混凝土组合梁

cavity wall filled with insulation 夹心墙

concrete small hollow block 混凝土小型空心砌块

control joint 控制缝

category of construction quality control 施工质量控制等级

composite subgrade, composite foundation 复合地基

characteristic value of subgrade bearing capacity 地基承载力特征值

cushion 换填垫层法

couposite rubber and steel support 橡胶支座

chord member 主管

crack resistance 抗裂度

concrete structure 混凝土结构

combined footing 联合基础

characteristic/nominal combination 标准组合

coefficient for combination value of actions 作用组合值系数

crane load 吊车荷载

characteristic value/nominal value 标准值

combination value 组合值

counting inspection 计数检验

control grade of construction quality 施工质量控制等级

cement deep mixing 水泥搅拌法

component 部件

core column 芯柱

common defect 一般缺陷

combined course 结合层

cold adhibiting method 冷粘法

cement flyash garvel pile 水泥粉煤灰、碎石桩

crosswise shrinkage crack 横向缩缝

construction joint 施工缝

cement-flyash-gravel pile 水泥粉煤灰碎石桩法

continuous seam 通缝

degree of freedom 自由度

design value of a load 荷载设计值

defect 缺陷

dimension lumber 规格材

damp 阻尼

dynamic moment of inertia 转动惯量

dynamic effect factor 动态作用系数

diameter of section 截面直径

eccentricity 偏心矩

design reference period 设计基准期

deep flexural member 深受弯构件

dynamic action 动态作用

design value of an action 作用设计值

design of building and civil engineering structures 工程结构设计

deterministic method 定值设计法

detail 细部

durability 耐久性

design situation 设计状况

design value of a property of a material 材料性能设计值

details of seismic design 抗震构造措施

displacement 位移

deflection 挠度

deformation 变形

dividing joint 分格法

ductile failure 延性破坏

dynamic coefficient 动力系数

design working life 设计使用年限

deep beam 深梁

dominant item 主控项目
dynamic consolidation foundation 强夯地基
dry jet mixing 粉体喷搅法
deep mixing 深层搅拌法
degree of reliability 可靠度
dynamic compaction, dynamic consolidation 强夯法
dynamic replacement 强夯置换法
design parameters of ground motion 设计地震动参数
design basic acceleration of ground motion 设计基本地震加速度
design characteristic period of ground motion 设计特征周期
dimension lumber 规格材(木材)
design working life 设计使用年限
design value of strength 强度设计值
equivalent uniform live load 等效均布荷载
earthquake 地震
earthquake focus 震源
earthquake epicentre 震中
epicentral distance 震中距
earthquake magnitude 地震震级
earthquake intensity 地震烈度
earthquake zone 地震区
earth pile 土挤密桩法
effective support length of earthquake action 地震作用
expansion and contraction joint 伸缩缝
elastic analysis scheme 弹性方案
equivalent slenderness ratio 换算长细比
effective width 有效宽度
effective width factor 有效宽度系数
effective length 计算长度
elevated overhead roof 架空屋面
effects of actions 作用效应
elastic deformation 弹性变形
earth pressure 土压力

explosion action 爆炸作用
element 构件
evidential testing 见证取样检测
fatigue capacity 疲劳承载能力
fundamental combination 基本组合
fundamental combination for action effects 作用效应基本组合
force (weight)density 重力密度
frequent value 频遇值
frequency 频率
floor live load; roof live load 楼面、屋面活荷载
first moment of area 截面面积矩
fatigue strength 疲劳强度
flexural strength 抗弯强度
frame 框架
foundation 基础
foundation soil, subgrade; subbase; ground 地基
fixed action 固定作用
free action 自由(可动)作用
frame structure 框架结构
folded-plate structure 折板结构
force per unit length 线分布力
force per unit area 面分布力
frame braced with strong bracing system 强支撑框架
frame braced with weak bracing system 弱支撑框架
flexural capacity 受弯承载能力
foundation 基础
force per unit volume 体分布力
fired common brick 烧结普通砖
fired perforated brick 烧结多孔砖
first order elastic analysis 一阶弹性分析
filler course 填充层
frequent combinations 频遇组合
full adhibiting method 满粘法
foundation earth layer 基土
flexible waterproof layer 柔性防水层
frame structure 框架结构

frame-shearwall structure 框架—剪力墙结构
frame-corewall structure 框架—核心筒结构
finger joints 指形接头
grouting foundation 注浆地基
ground treatment 基础处理
grout for concrete small hollow 混凝土砌块灌孔混凝土
general item 一般项目
glued lumber 胶合材
ground treatment 地基处理
glued laminated timber (Glulam) 层板胶合
gravity density, unit weight 重力密度(重度)
grade of waterproof 防水等级
geosyntheties foundation 土工合成材料地基
geosynthetics 土工合成材料
glued-laminated timber (Glulam) 层板胶合木
gap joint 间隙节点
handing over inspection 交接检验
heavy tamping foundation 重锤夯实地基
heat fusion method 热熔法
hot air welding method 热风焊接法
high-rise structure 高耸结构
height of section; depth of section 截面高度
impounded roof 蓄水屋面
imposed deformation 外加变形
inspection of structural performance 结构性能检验
intermediate assembled structure 中拼单元
inspection at original space 原位检测
inversion type roof 倒置式屋面
industrial building 工业建筑
inspection lot 检验批
inspection 检验
isolating course 隔离层
jet grouting 高压喷射注浆法
joists 搁栅
joint 节点
jet grouting foundation 高压喷射注浆地基

load effect 荷载效应
load effect combination 荷载效应组合
load-carrying capacity 承载能力
leaning column 摇摆柱
light wood frame construction 轻型木结构
log 原木
lime pile 石灰桩法
lime soil pile 灰土挤密桩法
length 长度
load effect 荷载效应
load combination 荷载组合
limiting design value 设计限值
load 荷载
laminated strand lumber (LSL) 层叠木片胶合木
laminated veneer lumber (LVL) 旋切板胶合木
liquefaction of saturated soil 砂土液化
lamination 层板
low frequency cyclic action 低周反复作用
linear strain 线应变
lengthwise shrinkage crack 纵向缩缝
life of water proof layer 防水层合理使用年限
limit states method 极限状态设计法
limit states 极限状态
limit state equation 极限状态方程
morter for concrete small hollow 混凝土砌块砌筑砂浆
mixed structure, hybrid structure 混合结构
mass density 质量密度
moment of momentum 动量矩
moment of force 力矩
masonry structure 砌体结构
steel structure 钢结构
member 构件
modulus of deformation 变形模量
modulus of elasticity 弹性模量
multiplanar joint 空间管节点
moisture content of wood 木材含水率

machine stress-rated lumber 木材机械分级
non-reinforced spread foundation 无筋扩展基础
normalized web slenderness 通用高厚比
nodal bracing force 支撑力
normal stress 正应力
normal force 轴向力
natural frequency 自振(固有)频率
natural period of vibration 自振周期
normal value of geometric parameter 几何参数标准值
oriented strand board (OSB) 定向木片板
overall stability 整体稳定
open caisson foundation 沉井基础
overlap joint 搭接节点
ordinary steel bar 普通钢筋
plank 板材
prestressing tendon 预应力钢筋
plain concrete structure 素混凝土结构
reinforced concrete structure 钢筋混凝土结构
prestressed concrete structure 预应力混凝土结构
pretensional prestressed concrete structure 先张法预应力混凝土结构
post tensioned prestressed concrete structure 后张法预应力混凝土结构
panel zone of column web 柱腹板支点域
prefabricated concrete structure 装配式混凝土结构
post-buckling strength of web plate 腹板屈服后强度
pile foundation 桩基础
pilastered wall 带壁柱墙
persistent situation 持久状况
permanent action 永久作用
permanent load 永久荷载
principal stress 主应力
prestress 预应力
principal strain 主应变
plastic deformation 塑形变形
partial safety factor 分项系数

partial safety factor for action 作用分项系数 pile 桩
probability of survival 可靠概率
probability of failure 失效概率
probabilistic method 概率设计法
permissible (allowable) stresses method 容许应力设计法
periodic vibration 周期振动
partial safety factor for resistance 抗力分项系数
partial safety factor for property of material 材料性能分项系数
period 周期
pneumatic structure 充气结构
perimeter of section 截面周长
polar second moment of area; polar moment of inertia 截面极惯性矩
plantied roof 种植屋面
primary structure 基体
parallel to grain 顺纹
perpendicular to grain 横纹
prefabricated wood I Joist 预制工字形木搁栅
parallel strand lumber (PSL) 平行木片胶合木
penetration 贯入度
preloading 预压法
prefabricated concrete structure 装配式结构
preloading foundation 预压地基
pressed pile by anchor rod 锚杆静压桩
primary linning 初期支护
Poisson ratio 泊松比
quantitative inspection 计量检验
quality of appearance 观感质量
quality of building engineering 建筑工程质量
quasi-permanent value 准永久值
quasi-permanent combinations 准永久组合
resonance 共振
radius of gyration 截面回转半径
relative eccentricity 偏心率
rise 矢高

resistance 抗力
repeated action; cyclic action 多次重复作用
restrained deformation 约束变形
retaining wall 挡土墙
rigid 刚架（刚构）
representative values of a load 荷载代表值
reference snow pressure 基本雪压
raft foundation 筏形基础
rock discontinuity structural plane 岩体结构面
reference wind pressure 基本风压
rigid foundation 刚性基础
retaining structure 支挡结构
reinforced masonry structure 配筋砌体结构
reinforced concrete masonry shear wall structure 配筋砌块砌体剪力墙结构
rigid analysis scheme 刚性方案
rigid-elastic analysis scheme 刚弹性方案
ratio of hight to sectional thickness of wall or column 砌体墙、柱高厚比
rigid transverse wall 刚性横墙
ring beam 圈梁
repair 返修
rework 返工
retention 保持量
rammed soil-cement pile 夯实水泥土桩法
reinforced masonry 配筋砌体
rigid waterproof layer 刚性防水层
reliability 可靠性
reliability index 可靠指标
span 跨度
shear wall of light wood frame construction 轻型木结构的剪力墙
sawn and round timber structures 普通木结构
stud 墙骨柱
site load 施工荷载
strength 强度
slenderness ratio 长细比

second order elastic analysis 二阶弹性分析
spherical steel bearing 球形钢支座
safety class 安全等级
standard frost penetration 标准冻深
snow load 雪荷载
subgrade, foundation soil 地基
structural concrete column 混凝土构造柱
shearwall structure 剪力墙结构
spread foundation 扩展地基
single footing 独立基础
strip foundation 条形基础
shell foundation 壳体基础
soil-rock composite subgrade 土岩组合地基
sawn lumber 锯材
section modulus 截面模量（抵抗矩）
tensile capacity 受拉承载能力
static analysis scheme of building 房屋静力计算方案
shell structure 壳体结构
space truss structure 空间网架结构
shear wall structure 剪力墙结构
shear force 剪力
stress 应力
section 截面
suspended structure 悬挂结构
simply supported beam 简支梁
superposed beam 叠合梁
sheet pile 板桩
slope protection; revetment 护坡
slab; plate 板
shell 壳
settlement joint 沉降缝
safety 安全性
strain 应变
shear stress; tangential stress 剪应力
self weight 自重
shear strain; tangential strain 剪应变

static action 静态作用
serviceability 适用性
shear strength 抗剪强度
shear modulus 剪变模量
shear capacity 受剪承载能力
stability 稳定性
spatial behaviour 空间工作性能
stiffness; rigidity 刚度
supposititious seam 假缝
serious defect 严重缺陷
set of high strength bolt 高强度螺栓连接副
slip coefficent of faying surface 抗滑移系数（螺栓连接）
soil-cement mixed pile foundation 水泥土搅拌桩地基
soil-lime compacted column 土与灰土挤密桩地基
space rigid unit 空间刚度单元
stud welding 焊钉（栓钉）焊接
sawn and round timber structures 方木和原木结构
step joints 齿连接
structural glued-laminated timber 胶合木结构
sampling inspection 抽样检验
sampling scheme 抽样方案
second moment of area; moment of inertia 截面惯性矩
story with outriggers and / or belt member 加强层
site acceptance 进场验收
seismic fortification measures 抗震措施
slab-column shearwall structure 板柱—剪力墙结构
structural composite lumber (SCL) 结构复合木材
sand-gravel pile 砂石桩法
seismic fortification intensity 抗震设防烈度
seismic fortification criterion 抗震设防标准
site 场地
seismic concept design of buildings 建筑抗震概念设计
studs 墙骨

structural wood-based pened 木基结构板材
structural plywood 结构胶合板
slab-column system 板柱结构
special engineering structure 特种工程结构
structure 结构
serviceability limit states 正常使用极限状态
shrinkage crack 缩缝
stretching crack 伸缝
spot adhibiting method 点粘法
strip adhibiting method 条粘法
self-adhibiting method 自粘法
silicification grouting 单液硅化法
soda soluting grouting 碱液法
shield tunneling method 盾构法隧道
surface course 面层
thrusted-expanded in column-hammer 柱锤冲扩桩法
transfer member 转换结构构件
transfer story 转换层
timber structure 木结构
troweling course 找平层
types of roof or floor structure 屋盖、楼盖类别
tall building 高层建筑
tube structure 筒体结构
tube in tube structure 筒中筒结构
torsional capacity 受扭承载能力
truss 桁架
transient situation 短暂状况
tensile strength 抗拉强度
thickness of section 截面厚度
tube structure 筒体结构
torque 扭矩
temperature action 温度作用
tributary area 从属面积
terrain roughness 地面粗糙度
timber structure 木结构
type inspection 型式检验
the smallest assembled rigid unit 小拼单元

test assembling 预拼装
ultimate strain 极限应变
ultimate deformation 极限变形
unbraced frame 无支撑纯框架
uniplanar joint 平面管节点
underground waterproof engineering 地下防水工程
ultimate strength method 破坏强度设计法
ultimate limit states 承载能力极限状态
under layer 垫层
upper flexible and lower rigid complex multistorey building 上柔下刚多层房屋
variable action 可变作用
vibration 振动
variable load 可变荷载

vacuum preloading 真空预压法
vibroflotation, vibro-replacement 振冲法
visually stress-graded lumber 木材木测分级
wind load 风荷载
wood-based structural-use panels 木基结构板材
wall beam 墙梁
wall 墙
wind vibration 风振
wood preservative 木材防护剂
wood-frame construction 轻型木结构
wood-based panel 木基复合板材
yield strength 屈服强度

Appendix 5 Bibliography 参考文献

I. Reference Books

1. 巴兰坦(Ballantyne, A), 王贵祥译. 建筑与文化. 北京: 外语教学与研究出版社, 2007
2. 蔡燕歆, 路秉杰, (新)李(Lee, A.).中国建筑艺术. 北京: 五洲传播出版社, 2006
3. 薄冰. 薄冰英语语法. 开明出版社, 2009
4. 薄冰, 何正安. 薄冰新编英语语法. 北京: 世界知识出版社, 2005
5. 卜德清. 中国古代建筑与近现代建筑. 天津: 天津大学出版社, 2000
6. 陈平. 外国建筑史. 南京: 东南大学出版社, 2006
7. 顾世民. 现代英语语法. 哈尔滨: 哈尔滨工业大学出版社, 2007
8. 何山. 影响世界遗产未解之谜的101件重大事件. 北京: 中国长安出版社, 2006
9. 《建筑创作》杂志社. 印象: 建筑师眼中的世界遗产. 北京: 机械工业出版社, 2005
10. 李观仪. 新编英语语法教程1. 上海: 上海外语教育出版社, 2008
11. 李观仪. 新编英语语法教程2. 上海: 上海外语教育出版社, 2008
12. 林立. 新编大学实用英语教程第一册. 北京: 对外经济贸易大学出版社, 2007
13. 牛津高阶英汉双解词典(第六版).商务印书馆, 牛津大学出版社, 2004
14. 钱永梅, 庞平. 土木工程专业英语(建筑工程方向). 北京: 化学工业出版社, 2004
15. 王受之. 世界现代建筑史. 北京: 中国建筑工业出版社, 1999
16. 吴承霞. 建筑工程专业英语. 北京: 北京大学出版社, 2009
17. 张道真. 实用英语语法练习与答案(修订版). 北京: 外语教学与研究出版社, 2000
18. 张道真. 张道真实用英语语法. 北京: 外语教学与研究出版社, 2004
19. 章振邦. 新编英语语法教程(第四版). 上海: 上海译文出版社, 2004

II. Reference Websites

1. http://anthropology.si.edu/maya/
2. http://baike.baidu.com/
3. http://baike.baidu.com/view/1663.htm
4. http://en.wikipedia.org
5. http://en.wikipedia.org/wiki/American_architecture
6. http://jan.ucc.nau.edu/~twp/architecture/
7. http://image.baidu.com/i?tn=baiduimage&ct
8. http://ngmchina.com.cn
9. http://solar.ofweek.com/2009-05/ART-260005-8220-28414199.html
10. http://www.artcn.cn
11. http://www.artcn.cn/Article/hysj/LLYJ/200605/9751.html
12. http:// www.bbker.com/viewimg
13. http://www.cctv.com
14. http://www.cctv.com/english/news/China/EduASciACul/20030804/100190.html
15. http://www.chinaculture.org/gb/en_travel/2003-09/24/content_33919.htm

16. http://www.chinapage.com/canal.html
17. http://www.costaricapages.com
18. http://www.doreandme.com
19. http://www.expo2010.cn
20. http://www.google.com/dictionary/
21. http://www.go-passport.grolier.com
22. http://www.jxenglish.com/
23. http://www.konarka.com/
24. http://www.mnitcc.blogspot.com
25. http://www.mwr.gov.cn/dlzz/english/waterinchina
26. http://www.nitchiewiki.com
27. http://www.si.edu/
28. http://www.si.edu/Encyclopedia_Si/nmnh/pyramid.htm
29. http://www.sunandclimate.com/facilities/list/105-photovoltaic-energy-in-missouri.html
30. http://www.topchinatravel.com/city_guide/Chengdu/attraction/Dujiang-dam.htm
31. http://www.51yala.com/html/2007822279-1.html